OUR UNIVERSE

An Alternate View

Damon Farris

ISBN-13: 978-1985700772
ISBN-10:1985700778
Copyright 2018
Damon Farris
Joshua Tree, CA 92252

All Rights Reserved

TOC

OUR UNIVERSE i

TOC iii

Acknowledgements vii

Dedication viii

Introduction x

Chapter 1 1

   In the Beginning 1

Chapter 2 10

   Many Mini-universes 10
   Dark Matter 13

Chapter 3 14

   First Stars 14
   Universe of Two Sizes 17

Chapter 4 19

   Receding Edge 19

Chapter 5 31

   Superforce At Work 31
   Distant Objects: How objects find each other 35
   Shadow Goes Global 40

Chapter 6 42

   Tides 42

Chapter 7 46

   Center of Transparency 46
   Bending Light Rays 48

Chapter 8 50
Darkening Shadow 50
Neutron Stars.. 54

Chapter 9 59
Powers of two in Operation... 59

Chapter 10 64
Doppler Effect 64

Chapter 11 70
Freefall vs Momentum vs Centrifugal Force 70

Chapter 12 73
Orbits 73

Chapter 13 83
Speed of light.. 83
Image of electron changing state.. 84

Chapter 14 88
Energy Requirement vs Light Speed 88

Chapter 15 95
Information Speed 95

Chapter 16 99
Space-Time Constant 99
A Sphere is a Sphere.. 105

Chapter 17 108
Time Dilation... 108

Chapter 18 .... **113**
    Flaws of Relativistic Mass 113
    Gravitational Problems 115
    Transparency Index vs Mass 118

Chapter 19 .... **119**
    Expansion 119

Chapter 20 .... **126**
    Ice Ages 126

Chapter 21 .... **147**
    Shockwave 147

Chapter 22 .... **152**
    Refraction—more factors of two 152

Summary **159**

Appendix A **163**
    Vectors 163

Glossary of Terms **166**

Index **172**

Acknowledgements

A huge thanks to our devoted proof readers:

        Richard A. Mitchell

        John E. Mitchell

And to our editor:

        Marion Vaughan

Dedication

In Memory of My Wonderful Sister
Margaret Louise Farris-Price

Introduction

When a storm cloud expands too much, its vapor changes form, cell by cell, to a more stable state—a liquid. All energy converts to matter at the same time. That is, all vapor within the cloud transforms into millions of raindrops in the same instant. This process emulates the procedure that brings forth space and matter: Water vapor simulates composite energy where each rain drop mimics each piece of matter. While rain turns into puddles, lakes, rivers and oceans, matter turns into horseshoes, horses, people and the universe.

This alternate view of the universe explains *why* and *how* the universe works as opposed to what it does: gravity for example. We take the position that gravity does not break away from composite energy. It remains behind and applies pressure to all matter. It does not attract, it drives objects together when a differential force occurs between two or more objects. Flux vectors reveal how that differential takes place. In other words, our Universe is pressurized, space is pressurized, and space is an entity that gathers all matter into ever-growing objects. And, like matter, it can neither be created nor destroyed.

We tell this story in present tense as we explain how the composition of super forces changes states from energy to matter leaving one of the four super forces behind.

The book is written at a high-school or drop-out level of understanding, so there is nothing complicated about how or why the universe does what it does. Any person interested in how length becomes shorter or how time passes slower with increasing speed toward that of light will know right away how that takes place. Any person interested in astronomy or how the universe works in general may be surprised at how simple this alternate view removes the mystery of why nothing can exceed the speed of light or even approach it in most instances.

Some readers may also be surprised to learn that the edge of their universe recedes at the speed of light, and it reveals new matter every day in the process. And another surprise may come when they realize how light travels on flux vectors. Yet, two more when they find out that their universe is only one of many and that space-time is a constant. Each person on earth lives within his or her own universe, and its edge always travels along with them. And they are always at its center. That leads to the title, *Our Universe: An Alternate View*

For those interested in global warming, you will find that Ice Ages are ongoing. They are either melting or freezing. The earth is near the end of its melting phase and will soon begin freezing in earnest. Within 6,000 years, there will be world-wide starvation because ice will cover most of the crop growing areas north of the equator, and there is not enough fertile land in the Southern Hemisphere to feed the world. That is, unless the Sahara Desert becomes productive again.

There are five (of many) important concepts that you will take away by reading *Our Universe: An Alternate View:* (1) the remaining superforce acts in a positive manner to crush objects—it does not attract, (2) new matter is revealed daily by the receding edge of our Universe at the speed of light; (3) if we could peer through the Cosmic Radiation Background, we would see composite energy or a Grand Energy from which we all came, (4) an object's properties does not change near the speed of light—a sphere remains a sphere; a box remains a box, and (5) failed subparticles result in dark matter.

But, the most important perception to come about is what the universe is expanding into. Hint: is not nothing.

Be curious—you're not that cat.

Chapter 1

In the Beginning

This is the story of our Universe: yours, mine, my friends', your friends'—all locally the same, all intermingled; where they all came from, how they all began, and how they merged to become ours—each having its own center; each having its own edge.

We are going to place myself at the starting line of the universe and speak from there as we watch it develop. Every now and then, we will pop back inside the real world when necessary.

In the beginning, only pure energy exists: a composite energy that contains all forces and all energies—a super energy. These forces are the strong interaction force, the weak interaction force, the electromagnetic force, and the gravitational force. Energies are many: thermal, kinetic, potential, and so forth. They are all wrapped up in one humongous sphere.

Now, let's repeat. In the beginning, only super energy exists. How that super energy or composite energy came to be is really unknown. It may very well have come about with the famous Big Bang. But our Universe does not begin there. It begins at the first appearance of space and matter. Space requires elaboration.

Space is probably the most misunderstood and misrepresented entity of our Universe. For this alternate view, it will be defined as an entity: a thing, an item, or an object. It is as much of an object as matter. Anything that occupies space is a thing. Anything that displaces it is matter; that is, only matter displaces space. Energy does not. It occupies space. Space is one of the super forces, a thing. To displace space means to push it aside and replace it with something else. To occupy it means to be inside and move freely about. Further, space has two properties, length and time. An object's velocity also affects each property, length and time, within its locality.

Space is created cell by cell—the leftover of what once was composite energy. The cells cannot be combined, but material and energy can move freely between them. They are elastic. They can flow and take on different shapes. Space is the summation of all those cells. And like matter, space can neither be created nor destroyed.

Right off the bat, we need to dispel the belief of some folks that space is a vacuum. It is not a vacuum, but it may contain a vacuum. People use the terms interchangeably, but they are not the same. Since we can't change past beliefs, we must interpret that when anyone discusses an atom and its orbitals operating in a vacuum, they really mean that the atom and its orbitals operate in space. Space separates an electron from its nucleus, not a vacuum.

In The Beginning

Trying to understand how an explosion brought about our Universe doesn't make sense. Explosions have centers where forces spread outwardly. Our Universe does not have one center. Everyplace is at its center. Even if one could travel from Earth to a planet a billion light years away in an instant, they would still be at the center; the edge would still be 13.9 billion years in any direction. Film at eleven.

What we're about to describe takes place in hyper-slow-motion. At this point, there is no metric for size or time. The references may be just tiny or large or fast or slow.

So, let us commence the story again from the beginning: the moment in time when only pure energy exists. If a Big Bang occurs, it has already happened. Professor Alan Guth refers to this time as Inflation. There are no laws governing light speed or anything else, so super energy spreads at a tremendous rate. It may be easier to break up the beginning in phases. Phase I is the Bang, phase II is composite energy, and phase III is the creation of space and matter. If that makes everyone happy, we'll begin there, phase III. Phase III happens, the others—maybe. If not, then we're at phase I.

Our Universe begins when composite energy expands enough to change states. We have no idea how large this volume of energy is, but it contains every force and every type of energy that will become part of our Universe. It could already be a quadrillion miles across, or it may already be billions of light years across. We don't know and don't care. We only care that it now exists. Super energy is expanding, volume of energy growing larger, density of energy becoming less—thinner. It is now so thin it becomes unstable: like a cloud of water vapor that becomes so cool it changes from a gas state to a liquid state, and all the vapor turns into millions of raindrops; all of them, all at the same time. Liquid displaces less volume than gas. Matter displaces less volume than composite energy, and matter displaces space; energy does not.

There is no temperature scale, no size scale. Only relative comparisons exist; hot or cold, big or little, fast or slow, thick or thin. Everything is contained in this super energy. There is no light, no heat. There's nothing else but super energy. Einstein's famous equation, $e=mc^2$, does not exist. And there are no nuclear explosions—nothing to explode.

Suddenly something happens. A small bundle of super energy precipitates. The bundle cannot sustain the same form when less force is applied throughout, so it condenses to a more stable state, matter. This condensation is matter in its simplest form, a filum. We use the Latin word for string to avoid confusion with *String Theory*. This filum is the smallest bit of matter to exist. Filums or filaments will go on to build elementary particles that go on to build electrons and protons. They are the building blocks of things to come—all things to come. It takes a huge volume of energy to make one tiny bit of matter, the larger the volume, the more the energy. Any form of matter is a reservoir containing all the energy transformed into its changed state. It will store that energy with little loss forever.

That little loss will come about by allowing some of it to escape slowly over time in various forms of radiation. Sometimes it will set its energy free in large quantities quickly when its state changes suddenly to another form. But that change will require tremendous energy to ignite.

It will take billions of years before scientists understand filaments, but for now, a filament is a step down from super energy. When super energy becomes less dense, or thinner, tiny filaments appear. In large numbers, filaments take on a different form or personality. They change as a crowd of peaceful individuals may change into a violent mob.

Imagine all the forces and energy folded into one tiny piece of matter. It leaves a large hole in the composite energy field. It is the smallest cell of space in existence, and the filum is the only solid piece of matter. Although the volume of energy to matter ratio is millions to one, the cell is smaller than a proton by far which means its filament of matter is a really tiny thing. The matter belongs to the space that created it.

Super energy, or SE, doesn't seem to get along with matter. A force pushes the filament to the center of the hole created by the change. This force also serves another purpose. It keeps the hole open. The surrounding SE cannot close in on the newly formed matter. If it did close in, there would be no space; it would be the end of our story. No universe would come into existence.

When SE becomes matter, that matter displaces a small volume that had been previously occupied by all those forces. However, not all super energy becomes matter. That force keeping the hole open is also brought to bear on the new matter. It wants to occupy the same expanse caused by the sudden displacement. This super force left behind is a flux. It is truly a force. It doesn't suck. It doesn't attract. It applies pressure to this new arrival. All lines of force want to be equal. Distance from the filament's ends to the boundaries of its hole is shorter than the distance from its sides. The shorter distance squeezes the flux like a spring. Flux doesn't like to be squeezed. It fights back. Since the force is greater on the ends of the filum, the additional pressure begins to roll the filament into a ball. Equilibrium is important to these lines of flux, and motion is as important to the filament.

In The Beginning

This single filament is running wild. It's alone and bouncing around in the huge hole left in place of its bundled energy. There is an invisible connection between the filament and the surrounding super energy field. The field recognizes the filament as ex-energy and repels it.

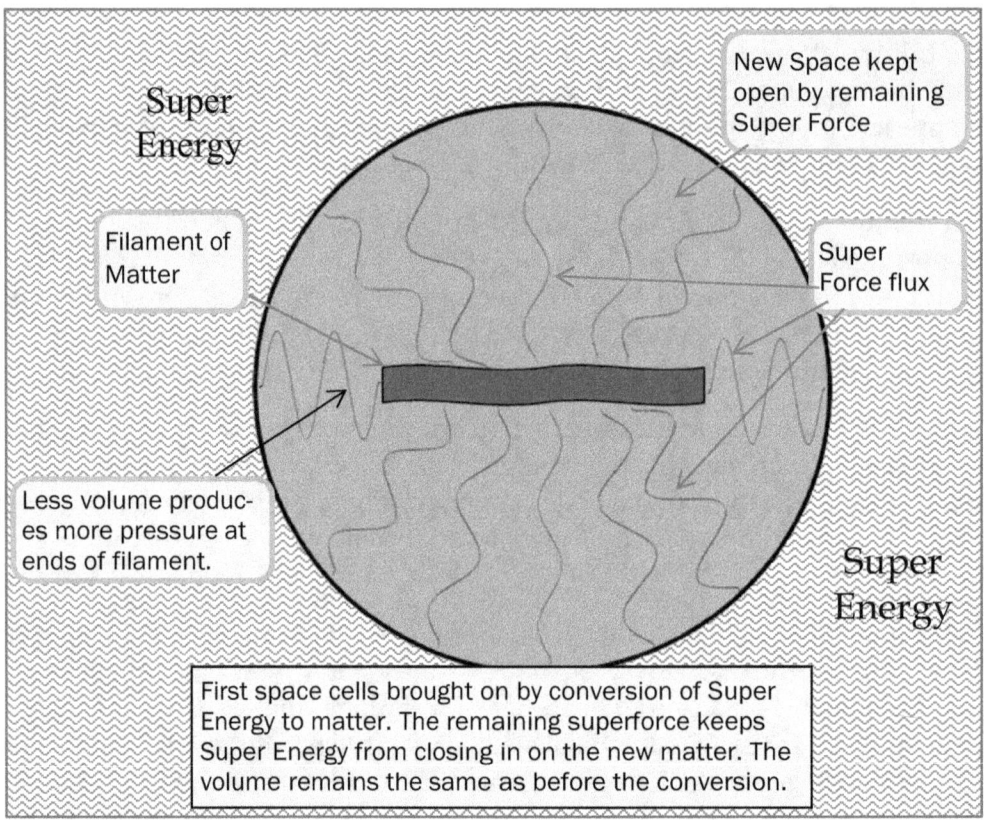

Another nearby bundle emerges and repeats the process, one more filament in its own extent. Two filaments running wild, fighting forces, crashing into the remaining field that defines their area and then getting knocked silly, whacked back to the center. The two holes leave a weak link separating them. The bond connecting the two holes breaks and is quickly absorbed by the surrounding field leaving one large spherical hole for two filaments to fight it out. They dislike each other. They repel each other, but the repulsion is weak. There is not enough room for the two to coexist, but the flux field of the remaining super force pushes them toward the center of the remaining space anyway. A fight with no winner continues as more filaments move into the crowded zone. But with each new filament comes additional volume temporally preventing filaments from becoming individual balls. Space is growing, becoming larger. Soon each small cell conjoins with its neighbor. Space is coming together in all local areas and all other areas of the composite energy—all in the same instant.

Filaments may come in several varieties with many characteristics. They tremble and move like a serpent or a fish. They send waves along their slim body from end to end. Some have a comfortable frequency they like—resonance. The bundled energy that produced the filament determines the frequency.

Many filaments have a single loop, some twist themselves into a corkscrew, and others wrap around themselves like jigsaw pieces. Something happens when they get within a certain distance of each other. It seems as if the end of one filament attracts the end of another. Two filaments with loops collide and become entangled like a Chinese puzzle. They are locked together and can only be undone by reversing the process, exactly. Others are mirror images. They collide and fit together perfectly like two magnets locked side by side—north to south, south to north. More and more collisions, more and more filaments combine and become balls: balls of filaments running around, growing with each additional filament pile on. Each ball begins to spin, slowly at first. As the balls grow, newcomers donate energy of motion to the ball causing it to spin faster.

Something weird is going on. The direction of spin depends on the view through the ball's axis, and the axis of each ball point in different directions. The combined forces of filaments making up the ball produce a charge whose polarity depends on the direction of spin. The direction of spin depends on the view—from above or below, from left or from right. But there is no up or down in space. Flip up-down ninety degrees left and up-down becomes left-right. The opposite side becomes right to left. Nothing is as it appears. Clockwise becomes counter-clockwise depending on position of view.

But charge is real and unmistakable. Suddenly two balls with unlike charges are drawn together disappearing in a flash. Nothing is left but a huge lump of energy dissipating throughout the former matters' remaining region. The change of state is very efficient. It left no matter behind. However, the state change is not efficient enough to get the matter back to its original state of super energy. Matter can never return to the composite energy from which it came: too many super forces were lost during the original transition. Another thing, the disappearing matter left its space intact.

We need a law that governs matter movement. That law will become the conservation of energy. For that, we need a new term: material, or better yet, quantity of material per object. We require another, one that includes speed and direction at the same time—velocity. Now we're cooking. Each filament has a certain velocity when it combines with another filament or a group of filaments or a ball.

Momentum, that's what we get when we multiply the material within an object, m, times the velocity, v. $M = mv$, that is what gives rise to spin. But we already have an M, so it gets another letter, ρ—the Greek letter rho. The momentum of all filaments summed together using their direction produces the spinning ball. It will be either clockwise, or counter-clockwise, or left to right, or right to left.

When a ball grows to a size the new universe deems stable, it stops growing. It's no longer just a ball. It is a lepton. Soon there are quadrillions of leptons. They take on their fundamental material's personalities, except in a highly amplified way.

That slight repulsion? Forget about that. When all the filaments get together, that slight becomes gigantic. Leptons hate each other. They refuse to stay close and the field of energy that kept them close before is overwhelmed. They bang against the outer edge of their space avoiding each other. Super energy fills that position and knocks them away. The fight literally heats up. Now there's HEAT, and it's getting hotter, much, much hotter. Leptons in motion emit electromagnetic waves. Now there is LIGHT. Some of it, along with lepton radiation and opposite charge collisions (matter-antimatter), will be left over and discovered billions of years from now to become known as Cosmic Radiation Background.

Nature is not done with these new particles yet. Although stable, they are not as secure as the universe requires them to be. They want to live a long life, but some have too many filaments attached at one location. They wobble as they spin. They need more balance to survive. Nature can fix them. The universe demands these particles be perfect, reliable. The unwanted filaments must be removed, but they are locked in place. How?

Turn them back to energy. Not the super energy they once were, that's forbidden, but to some lower grade energy. Anything will do. Just get them off this rotating body so it doesn't wobble enough to disintegrate. These unwanted filaments want to hang on, but those opposing outnumber them. Another fight, an unwanted filament is flung off at a terrific rate—thrown back into space it helped create. But it's free, and it's weightless. It's energy again and headed off to whom knows where. However, the body had to spend energy getting the thing off. It was just enough to set the lonely one on a journey to another state change. The journey will be known as radiation at some future date. The lepton continues to lose unwarranted material in this manner until it reaches a perfect state of balance and charge. Finally, it becomes completely stable—it is an electron. It will remain so forever or until some tremendous force acts on it.

Radiation will always be filaments way of converting themselves back to energy. Not the original it came from but energy none-the-less. It's a two way street. When an electron needs material, it snatches a filament's worth of energy from its surroundings, and the filum changes its state back to a solid that becomes part of the electron.

The conversation is restricted to the more familiar basic subatomic particles, so when an electron appears, recognize that it may have altered its appearance several times since it changed from the pure energy state. Further, the use of the word material refers to some number of filaments.

Although the universe is growing, there is limited room. The remaining super force squeezes all matter together. But electrons still don't like it, so repulsive actions continue. They attempt to keep everything apart. Electrons repel other electrons, super energy repels matter, and lines of flux want to gather everything to a central location. This repulsion manifests itself into an outward pressure that guarantees a sphere of space surrounding all matter within. What a mess.

When a portion of super energy goes through the process of becoming an electron, that electron inherits all other links to all other matter that has undergone a change. However, the link's role has been reversed. The new electron knows it is ex-energy and must abide by certain unwritten laws of nature that make it an electron: it must repel any other electron, it must produce a magnetic field perpendicular to its direction of travel, it must gain material while absorbing electromagnetic energy, and it must lose material while releasing electromagnetic energy.

Soon, strangers appear. They are not electrons. It seems a few filaments had gotten together and made themselves into a cookie-cutter. The result appears to be mass production of the strangers—more elementary particles. They are quarks, all types of quarks for future atom construction. Quarks are complex, and they will go on to build particles that are even more complex. In the future scientists will put leptons and quarks in a class called fermions.

Other strangers appear. Somehow, other matter governs how the complex quarks operate. They are various bosons that act as mediators—bosses directing what to do, how to do it, and when to do it. When filaments go on to build large complex elementary particles, it appears as if forms and jigs come into play that act as patterns for production of behavior, a precursor to DNA (Deoxyribonucleic Acid) if you will. They may control two super forces of composite energy, that is, the strong force and the weak force: the forces that allow construction of atomic nuclei. The more complex subparticles become, the more complex their actions become. After all, without proper behavior of particles that make it up, no ordered universe could exist.

Our Universe is born through expansion, and it continues after birth. Remember, expanding composite energy brings about matter. Billions and then trillions of particles create even more nodes of space, and these spheres cluster and spread throughout the new universe. They collide, maintain contact, and space and matter appear in one super large sphere. And it keeps on growing. Each nodule acts like a spring. When thinning super energy creates another node, it is already in motion.

In The Beginning

Far into the future, a creature will interpret this motion as if our Universe is expanding at the speed of light. But it is not. Revealing itself at the speed of light, it is. Our Universe has more matter today than yesterday because light revealed more of the universe in the past twenty-four hours. The edge is 16,094,799,101 miles farther away in every direction. In a few million more years, the new matter revealed today will have become quasars or galaxies along with an additional 84% more space. How that extra space appears will become clear soon.

With each new quark comes even more space. Quarks do not fight. They get along just fine and leave a greater distance for electrons to separate farther away from each other. As more quarks come into existence, another subparticle appears. Somehow, the bosons have rendered themselves even more helpful. They are gluons. Their job is to regulate quarks in such a way as to hold them together while building up a very complex, fully grown subatomic particle.

As the universe grows, DISTANCE between warring factions increases, and things settle down giving a cooling off period. Yes, since there's heat, there is cool.

There is now space, distance, light, and heat. With distance comes a need for one action to influence another over that interval between. How far away? How long it takes? TIME. Add time to our list. Further, combine two properties of the universe into one, space-time. One of this author's favorite scientists sometime refers to the two as the *fabric* of space-time. In a way this fits where flux represents thread, waft represents length, and warp represents time. A very important and necessary property of matter is still missing. It already exists but not defined. That definition will have to wait.

When energy changes to matter, the amount of energy is in direct proportion to the material making up that matter. At this point, imagine the amount of material being a number of filaments. Of course, that number is TBD (to be determined.) As in the beginning, volume of super energy is a representation of that material. That is, particles with the most matter require more volume. A standard relationship of that volume of energy will come about in a few billion years that will depend on its type. Some quantity of electron volts per unit of material will define that relationship.

While leptons are becoming stable as electrons, quarks and bosons and gluons are busy forming up to become a much more complex object, a proton. Its charge is opposite than that of an electron; it is positive with a sign of (+). Of course the opposite of plus is negative with a sign of (−). Protons repel each other just like electrons, and they have much more material—over 1836 times more material than electrons.

As holes grow when more matter forms, a greater volume becomes available as they unite. Electrons find friends in protons. They are drawn to each other. Since the proton is much larger and heavier, the electron does the traveling. When it approaches the proton indirectly at a great pace, it misses but continues to circle the proton. When it finally settles down in a stable orbit, together they become neutrally charged. Nature prefers neutrality, stability. The new pair is magic, an atom. The proton becomes the nucleus. Together they will be called hydrogen and be known as an element: the first and tiniest element of all yet to come. When another electron approaches the pair, the one in orbit repels the new guy.

If an electron approaches a proton at a bad angle, it collides with the proton and releases all of its energy of motion. The added material forms another subatomic particle. It is totally neutral because the charges cancel. The new sub particle is a neutron, and it roams around the growing universe without a care. Collisions between electrons and protons occur less frequently than the capture of an electron by a proton, so protons outnumber neutrons. However, a lone neutron is short lived. If it doesn't join up with an atom right away, the electron is cast off along with a weird object generated from filums of the neutron, a neutrino.

Since electrons and protons have no effect on the neutral object, the superforce has its way with a neutron. It can combine with a proton by giving up a little of its energy. Hydrogen is comfortable with its new friend and together they make an isotope, a very stable isotope. While not a new element, it provides a new attribute to hydrogen and will be become known as deuterium. Rarely, a second neutron will join the other to become tritium.

Hydrogen atoms are still single as our Universe continues to make up rules. A lone electron going around a nucleus is deemed unstable. There is another new law, and it has a mathematical relationship. The minimum number of electrons allowed to occupy an orbit is two, and that orbit's identifier depends on its distance from its nucleus. It begins at two electrons, and it will go up from there depending on future's nucleus' makeup.

But for now, hydrogen has no other choice. The separation is too great for individual atoms to be concerned. However, a few may come into contact and form up as a pair to become a very stable molecule of hydrogen. That is two electrons in orbit around the nucleus of two protons. Sometimes a neutron will accompany the protons and become part of the atom. Our new Universe is coming together in a hurry.

Chapter 2

Many Mini-universes

The forgoing is taking place at the same instant duplicate processes are underway. Super energy is turning into matter throughout the volume of this Composite Energy, and the universe is already billions of light years across before the startup processes are done. Finally, there is little Composite Energy remaining in this part of the universe after all cells have come together. The lingering globs of super energy maintain their characteristics. They still repel matter when it comes near. Such globs may become mysterious to scientists billions of years from now as dark energy.

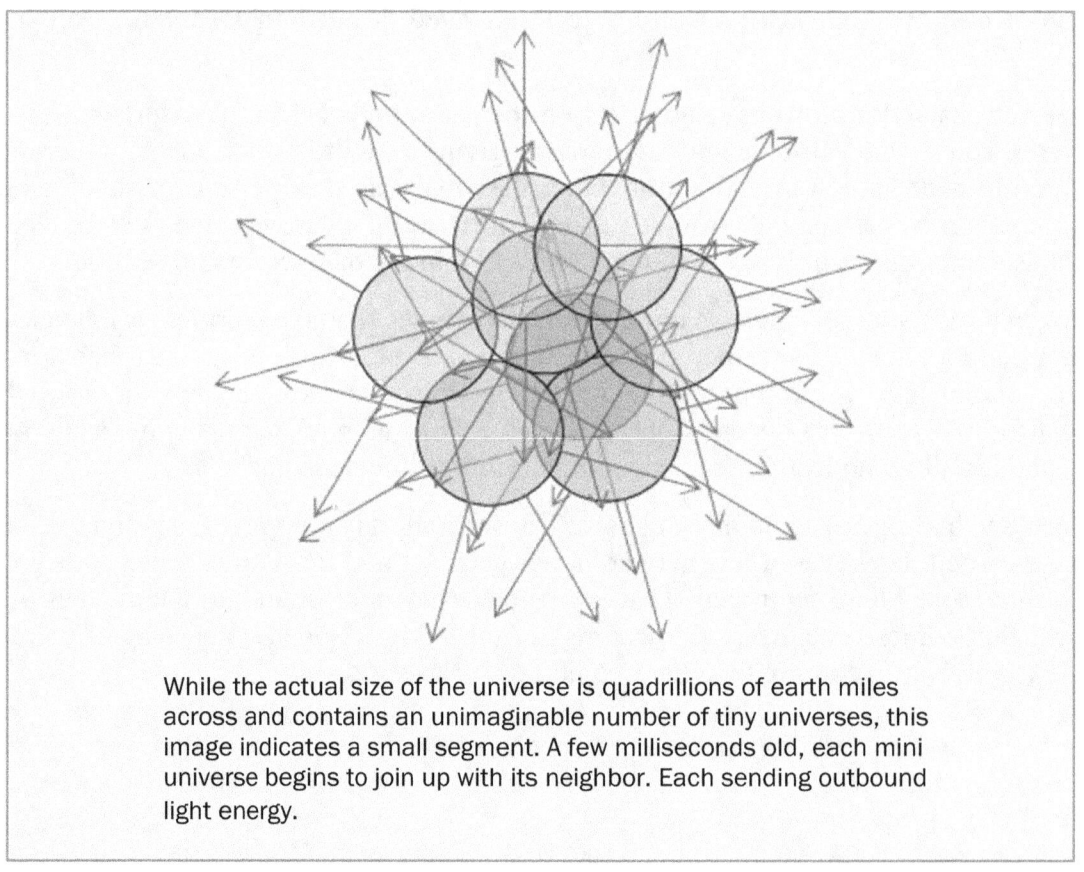

While the actual size of the universe is quadrillions of earth miles across and contains an unimaginable number of tiny universes, this image indicates a small segment. A few milliseconds old, each mini universe begins to join up with its neighbor. Each sending outbound light energy.

Building the atom comes about as each cell merges with its neighbor and finds its new friend beneficial. When it does, it presents an opportunity to begin a foundation of something greater than the two separately. Then as that combination grows, and more filaments create other products with each new merge, they become more useful. When it finds its new friend detrimental, that opportunity never occurs, and failed subparticles are born.

Soon we have a complete object, maybe two, which join up. Perhaps a full-fledged quark meets another, and then they meet another with the right qualifications to suggest a special behavior. A proton is born. Later an electron's universe expands into the proton area, and more magic comes about—an atom, and the beginning of something very special.

The image below is a representation of the universe at only a few seconds old.

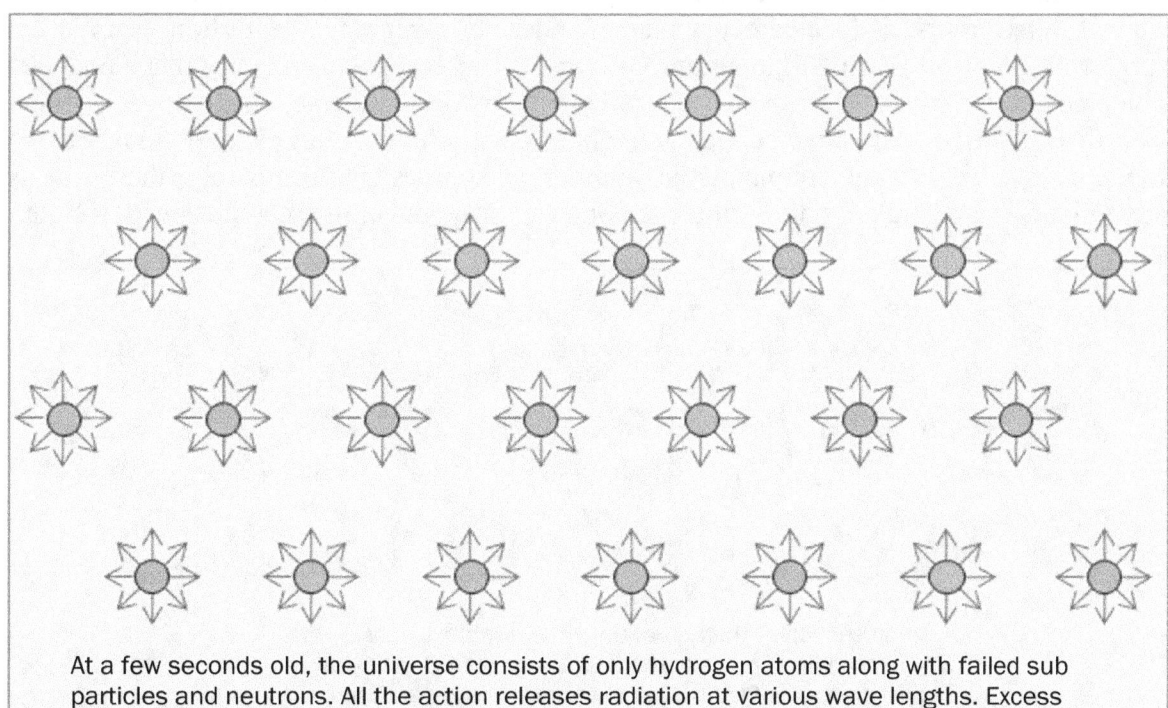

At a few seconds old, the universe consists of only hydrogen atoms along with failed sub particles and neutrons. All the action releases radiation at various wave lengths. Excess space separates the mini-universes. The excess was brought on by failed sub particles and disintegration of matter through particle/antiparticle annihilation. At ten seconds, each universe is 6,000,000 kilometers across.

In The Beginning

At 300,000 years old, things begin to settle down. While really unknown, the universe could be as large as 150 billion light years across at only a few years old. The image below represents a small section of the universe at that young age. Each mini-universe has chowed down on its neighbors to reach the size. And each one continues scarfing.

In spite of space growing at 5 ¼ times the rate it should be, clumps of matter still cling to each other forming clusters. Soon they integrate into even larger groups which quickly become the whole.

Understanding how each universe gobbles up its neighbor is essential to grasping how the center of our Universe is located every place inside. For example, 13.9 billion years in the future every place on Earth will be at our Universe's center. However, every place on Mars will also be at its center. One reason the reference is to "our Universe," is that every person on Earth is at his or her universe's center. That is, the universe begins everyplace at the same time resulting in an unimaginable number of sources. Unfortunately, there are as many universes as there are points, and each one can only reveal itself at the speed of light.

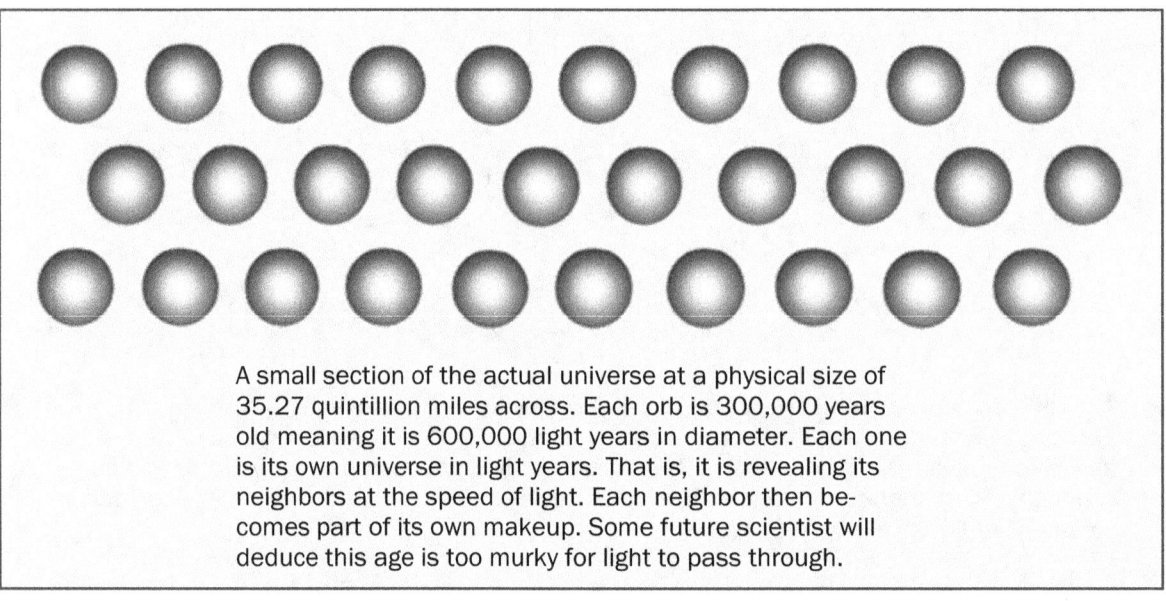

A small section of the actual universe at a physical size of 35.27 quintillion miles across. Each orb is 300,000 years old meaning it is 600,000 light years in diameter. Each one is its own universe in light years. That is, it is revealing its neighbors at the speed of light. Each neighbor then becomes part of its own makeup. Some future scientist will deduce this age is too murky for light to pass through.

A mental conflict between physical distance and time distance arose when presenting the physical size of the universe and the size of the mini-universes or orbs. All of the newborns are 300,000 years old, so they are aging in parallel. Regardless of how many exist, they are still the same age, so the distance across all of the universes during that period should not be measured by time. Instead, a physical size should be calculated. It turns out to be over 35.27 quintillion statute miles across for these ten babies. One could probably push it and convert that distance to light years, but then that size could be misunderstood to be x years old. See what we mean about conflict?

Dark Matter

Not all filaments combine into elementary particles. They do not fit together properly to make up anything worthy of working with others. They are failed fundamental particles that form up in large groups, lumps of filaments if you will, that lie around and do nothing. Yet they will influence our Universe. These failed bits of subparticles will be known as dark matter in the far future.

In 84% of its attempts, our young Universe fails in manufacturing viable products. That number includes matter-antimatter collisions which act as if that matter just up and disappears. Imagine this. Only 16% of composite energy makes its way to becoming worthwhile particles. That means a lot of extra space, brought on by failed filums at making something useful, can be used to separate meaningful matter products. The separation exceeds the remaining super force's ability to gather all matter together at the universe's center point. Such an enormous distance between large groups of particles reduces the force of flux, so it has no power over the whole. Its power influences local groups only. That extra space also allows inference of an expanding universe. The positive side is that a lot of energy reservoirs are hanging around as dark matter assisting in the makeup and control of yet-to-be colossal galaxies.

This is another imagination exercise. Suppose that all attempts of changing composite energy to meaningful matter were successful. That means that large voids would never exist. That means that giant galaxies and other stuff would fill these holes. That means that future intelligent beings would never be able to see through these bright balls of matter to discover such a large universe. It would be interpreted as being only a few million light years in depth, and along with local stars and galaxies, the night sky would be glowing with unseen objects—just unresolved light beams scurrying across the sky.

And, it means that the remaining super force could gather all matter together and crush it into one super-duper black hole. Perhaps to be recycled someday, but matter would still be gone resulting in a failed universe. It may have happened before.

Chapter 3

First Stars

Some TV programs tell us about the formation of the first stars. When the narrator says they are made of gas and dust, I want to scream out. Throw the remote at the screen. There is no dust. Dust comes after many generations of stars.

So, here's an alternate view.

Sometime after hydrogen atoms are built, including isotopes, the remaining super force gathers all of the atoms within its range. Remember the extra room between large groups left over from failed subparticles? Well, that extra space comes into play immediately. It separates hydrogen atoms into clusters such that they can be gathered together as groups instead of one large collection that fills the young universe: yea, what a star that would have been.

These clouds are the largest assemblies of atoms ever to be collected, or ever will be collected. Not millions of miles across but billions of miles across. The atoms have no charge. They are neutral. They don't care if they are close together. So, in a short time the remaining super force compresses these atoms closer and closer together. Most of them become partners. They are couples, the second generation of molecules.

At first, there is little resistance. But after joining to become molecules, they reach a limit. They begin to push back. The pushback is in vain because the remaining super force is the most powerful compressing force in existence. Hydrogen can only take so much togetherness. Molecules like their freedom. Banging into one another is not freedom. They grow hot. The superforce isn't finished. It brings to bear more and more pressure. Hydrogen becomes hotter and hotter. Our Universe has declared that a molecule has its freedom, and it will defend it with heat energy. The closer atoms come to each other the more heat energy they release. But the energy has no place to go except to its neighbor. Transferring the heat is akin to a bucket brigade fighting a fire. But the buckets are empty of anything but flames.

At a certain stage, physical compression comes to a critical point. The molecules are as close as they can get and remain in their natural atomic configurations, so the electrons begin to change orbits.

When hydrogen molecules are pressed together beyond this critical limit, adjacent atom electron/electron action forces the orbitals closer to the nucleus. The universe has also amended laws. Electron orbitals are restricted to certain levels, and when they change orbitals, there is a price to pay. When moving inwardly they must become lighter; that is, they shall release some material. As said before, the material released must become energy. The energy is in the form of radiation. The amount depends on how far the electron moves inwardly. While hydrogen has only one electron, it must still abide by those rules set forth and yield energy at every level.

Let's side step for a moment. This experiment requires a rocket and a payload to put in orbit. It's a simple rocket and payload, something like Sputnik I, the first satellite Russia put into Earth orbit, October, 1957. The idea here is to realize the difference in energy required to launch the same payload to various heights. It's easy to understand that to put a satellite in a 400 mile orbit requires more rocket fuel than it does to put the same one in orbit at 200 miles. And even much more fuel, therefore energy, to place the same object in a 1000 mile orbit. Not all the energy goes into the satellite. Most goes into the transport vehicle and its fuel just to get it into orbit, so very little gets into the satellite itself. Still, it has more energy in a high orbit than in a low orbit. So, the higher the orbit, the more the energy required to put it there. If the objective is to have the satellite escape Earth's gravity, that requires even more energy. Think, Voyager or Mars Lander.

Then there are asteroids, those asteroids running around the sun between Mars and Jupiter. While wandering round and round the sun, they seem harmless enough. Yet they have a tremendous amount of energy. It is potential energy. A large rock hitting the earth from the asteroid belt inflects much more damage than the same size rock from a 200 mile orbit.

Such is the electron. An electron has more material and therefore the most energy when it's free, running wild. When a hydrogen nucleus captures it, it must lose matter in the form of energy. That is, a filament is cast off to return to energy, maybe more than one filum depending on the quantity of energy required to do the job.

So why does nature require the electron to dispel or absorb matter when it changes orbits?

As before, the explanation requires using a satellite. Suppose a powered satellite moves from a 200 mile orbit to a 250 mile orbit. The satellite first must gain speed to get up there but slows down when it reaches the higher orbit. Its new orbit is not circular. It is elliptical and at its apogee—its highest point and slowest point. From there it falls towards Earth and gains speed until it reaches its perigee, its closest point and fastest point. Then it trades speed for altitude climbing back up to its new height. Certain other maneuvers are required to put the satellite into a circular orbit from there. Further, if the satellite is huge and weighs tons it only requires more fuel to get in its position. The orbit will remain the same.

Here's the strange thing about orbits under the influence of the superforce. Once stable, weight of the object has nothing to do with its distance from Earth. It could be as heavy as an M1A1 tank or as light as a feather. It doesn't matter. The only requirement is its speed and location above the center of the earth.

Putting It Together

While asteroids revolve around the sun instead of the earth, the law is the same. Some objects are hundreds of miles in length—a few spherical—while others are the size of a grain of sand. The material in each one varies by millions of kilograms, but they all remain in the same orbit. That is until something influences the object. Think dinosaur extinction. Further understanding of why this is so will come during the subject of freefall.

However, electron orbitals are not under the influence of gravity. They must obey different laws: laws that govern atoms, laws that subparticles having charges must obey. An electron's speed does not determine its orbital. The speed remains the same no matter what the distance is from its nucleus. However, the farther away from its nucleus, the less influence the mutual attraction becomes. Under these terms, it's rather easy to understand that an object closer in to its nucleus having more material as one revolving around at the same speed as one farther out, will not remain there long. It will be sent flying out so far as to become detached from its nucleus. But, if that same object becomes much lighter, it could remain in the required orbital forever, or until something forces it to change. Since electrons orbit at the same speed, about 90% c, they must release matter in the form of energy or gain matter in the form of energy when they change orbits.

So, the first stars have two things going for them to ignite: heat due to compression, and heat due to radiation. Because of neighboring electron/electron reaction, when atoms get squeezed too close together the outer electrons are forced to give up their orbits in exchange of energy for inward orbits. And heat due to radiation is much more than that caused by compression, thousands of times more. It's akin to atomic bombs. Material turns into energy. Except, electrons can recover the material, atomic bombs cannot.

Once the heat and closeness reaches a critical point, nuclear fusion begins. The two atoms of a hydrogen molecule become one atom of helium releasing a tremendous amount of energy. Things become even hotter creating more helium. But these super large stars don't last long. They blow themselves up creating heavier atoms. Finally, after some time and several generations of stars, dust in the form of silica based molecules and other elements come into existence. Now comes the dust. It becomes scattered about, intermingling with hydrogen gas and what-all.

The subgroups of hydrogen are so large that one super star becomes king of the isolated group. It alone contains most of the hydrogen, but after the first detonation, its children fill the same space. Over time, they make several huge stars and generate more new elements, but the available space remains the same. Nothing new is created because that process requires transformation of composite energy to matter, and that's all used up in this area.

What was once just a large group of hydrogen atoms is now a large group of stars. After some time, many grandchildren of the superstar exist, and then sometime later, a little dark matter is thrown into the mix.

Something is still lacking. A new metric is needed to realize the size of these first groups of stars. It will be known as the distance light travels in one year. Currently the star count is in the billions, but they wander aimlessly—spread apart too far for the remaining super force to have much of an effect. They are useless for the further advancement of our Universe until all those failed subparticles act in unison.

Remember those first filaments of matter are solid even though they are tiny. Solid objects have a transparency index of zero. More on this later, but for now realize that no superflux gets through a solid. A differential force gathers locally available failed subparticles together into the heaviest object to ever exist. It will be known as a black hole. Over a short period of time, superflux forces the supercluster of stars around this black hole, and it works magic. This supersized object of total blackness influences stars near the center. They revolve around it at a high rate while stars farther out tend to form a disk and revolve in a stable orbit.

However, stars, gas, and dust created from grandparents that are farther away from the center at the outer edge seem unstable. There is not enough material available to keep them attached to the group: more dark matter to the rescue. This seemly worthless failed matter has found another use. It weaves itself throughout the outer region and reigns in wayward stars and even whole solar systems. Soon a beautiful spiral galaxy forms. But the process has been going on in parallel with billions of other galaxies. Smaller, larger, and of various configurations, and our Universe is just a few million Earth years old.

Universe of Two Sizes

For our purpose, there are two methods for measuring the distance across the universe: the actual size of the entire collection, and the size derived by light.

Let's look at its actual size first. At one year old, the universe collective is already billions of light years across because that's how large the composite energy source was when it had expanded enough to bring space and matter into existence. For a detailed explanation of the observable universe, see https://en.wikipedia.org/wiki/Observable_universe because that discussion is far beyond the scope of this work.

The same conflict presented earlier grows when the current size of our Universe comes into play. When referring to the collection of all universes, the size in light years is the total of all those universes added together. So, when someone says the physical universe is 150 billion light years across, we can assume they are talking about the collective.

Revelation at speed of light

With all that under our belt, the size derived by light is determined by its speed; number of universes is determined by the observer's location. Our Universe's size is derived by light, and it has grown from a few nanoseconds wide to 27.8 billion light years wide.

The universe is expanding at the same speed it was when matter came into existence. That is, when SE expanded and cooled enough to give us matter. It also stands to reason that anything under pressure, if not restricted, will expand. An example is a rubber balloon filled with air. It will either expand until the rubber exerts an equal pressure to counter the internal pressure or burst.

Since the universe is pressurized, it must also expand. But what is it expanding into? It cannot expand into space because that is part of the universe, but whatever it is, the universe is replacing it with itself. If nothing surrounds the universe, we must ask ourselves, "What is nothing?" The real definition even boggles the minds of a great many scientists. No matter what noun one can imagine, vacuum, space, or any other name, it is still something.

The edge of the universe marks the beginning of our Universe and the end of composite energy. There is a marker for that boundary. It's the CRB. One side is composite energy, the other side, our Universe. Even though the only remaining superforce pressurizes the universe, composite energy contains that same force along with all others. Therefore, composite energy is the arresting power that restricts a runaway condition of expansion, and we can say that the universe is expanding into composite energy. That is, the edge moves into the area as composite energy generates space and matter. Hopefully it will all become more clear while demonstrating the receding edge.

Chapter 4

Receding Edge

The collective universe is only a few million years old but is already billions of light years across. However, if intelligent beings have developed on some young planet, they will only see the universe as a few million light years across. That is because superflux carries light rays at a given speed, and the source comes from everywhere. Since superflux carries light at a given speed, our Universe can only reveal itself at the speed of that flux. To each of those intelligent beings on the young planet their universe is a few million light years across. And they assume it is only a few million years young.

Receding Edge I

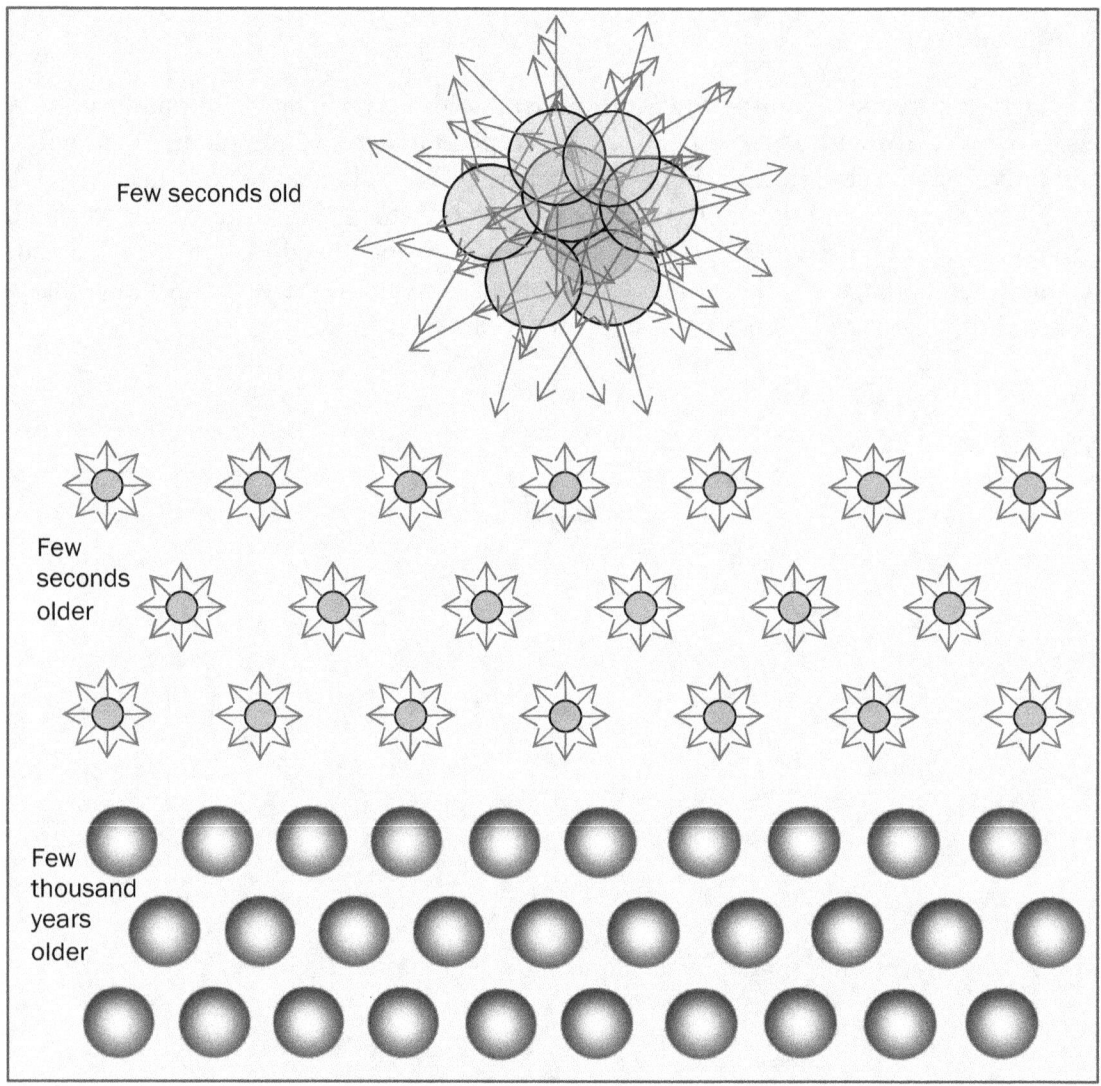

The images above have been reproduced as a reminder that the edge of each mini universe recedes, scarfs up its neighbor, and continues to gain matter at the speed of light.

Let us begin with the middle group, the image of a few seconds older. We'll set up a matrix to identify each mini but concentrate on only two, E4 and I4.

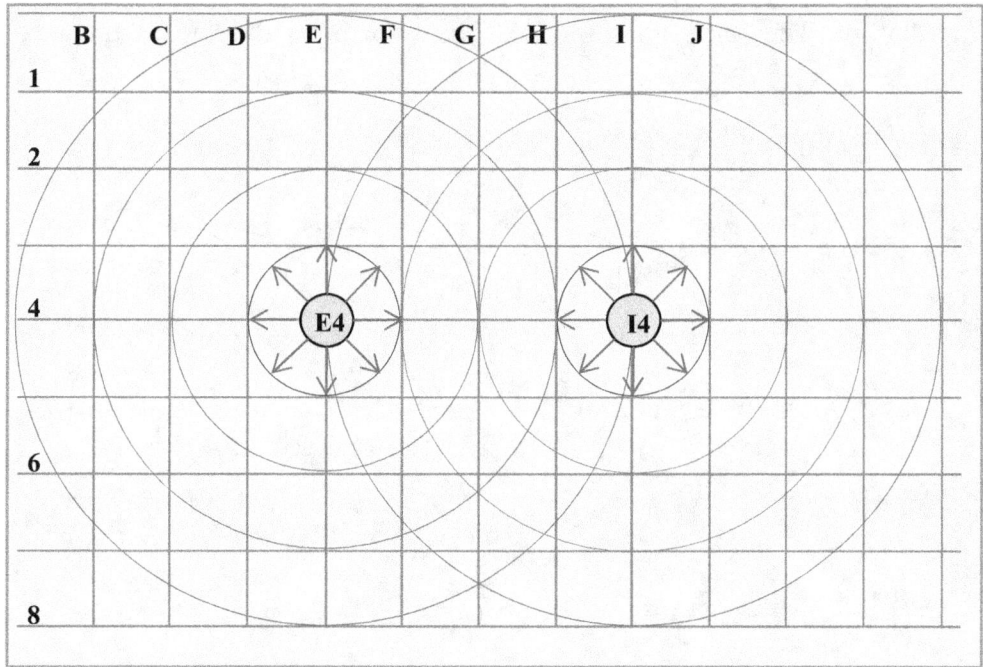

The grid is a matrix of one light second increments.

Looking at mini-universes E4 and I4 one second after space and matter creation, tells us that their information has traveled one light second in every direction. That is, these two mini-universes are one second old along with an uncomfortably large number of others. There are many neighbors between these two, but they are not shown for simplicity. This matrix shows how each universe intrudes on its companion to become conjoined, and then the two become one, and it shows how the edge of its universe recedes—all of which happens at the speed of light.

While radiation is in all directions, let's concentrate on directions left and right.

After two seconds, their light rays meet at intersection G4. After three seconds, each has entered and mingled with its neighbor at F4 and H4 after joining with all others between, but they have not joined each other yet, just inter-mingling. At four seconds, E4's light has entered into the starting point of I4, and I4's light has entered into the birthing point of E4. Now a portion of eastern E4 is part of I4 and the western part of I4 becomes part of E4.

E4 has aged four seconds when its light arrives at I4, but information in that light says E4 was just born. I4's edge of its universe has receded four seconds and now claims all the matter in that portion of E4, plus all other universes it has gathered along the way, plus what it had before. The same goes for E4. The two minis overlap as the next graphic shows.

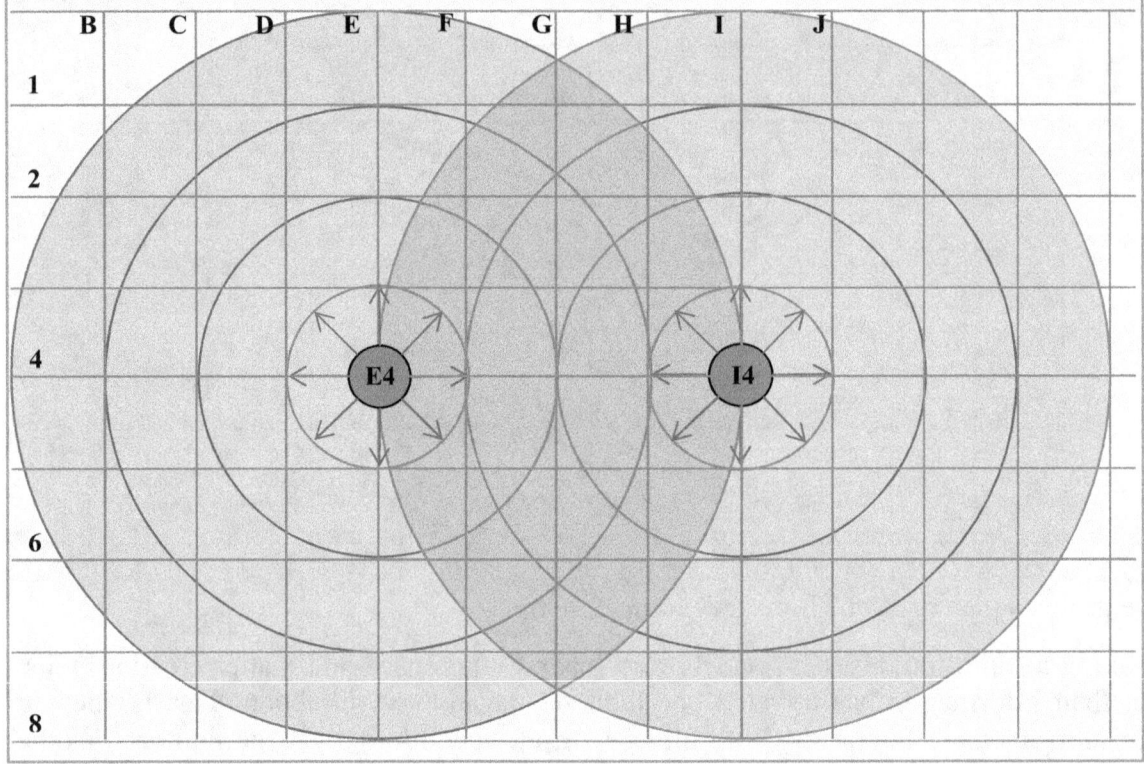

The darker area indicates matter shared between the previous two universes. We are looking at a plane cut through the globes of the two overlying cosmoses.

Let's take this idea a little further. Suppose that instead of seconds, the distance indicated by each square is two billion years. Eight billion years gives enough time for a new E4 and a new I4 to be located in two separate solar systems with intelligent life. Each system is located at the center of its universe.

If space travel has been invented at wormhole speed, a traveler could leave E4 and stop at I4 and still be in the center of the universe. Specifically, the center of their universe moves along with the traveler. That is because the person is always equal distance from the edge with respect to light speed and distance.

As she moves toward the edge, she passes through all the center points of all other universes along the way. Some points may be in deep space while others may be in the center of a rock or star. Recollect that each universe begins at a point where space and matter were created and spreads outwardly at the speed of light, and there is much more space than matter.

The following is relative to J9, but each point can tell the same story.

When E9, J13, and N9's information reaches the birth point of J9, that information is in the form of CRB. J9 can look back at the birth of each one. Three seconds later, information arrives from B9. The edge of J9 is now seven seconds away at D4, B9, and J16. Now, E9, J13, and N9 are three seconds old as far as J9 is concerned, and J9 has aged seven seconds. Yes, to some that may sound crazy, but only three seconds have passed since discovery of E9. Not included are the four seconds away it really is. When we include the four second distance to E9 and the three second distance to B9 together, they make J9 seven seconds old.

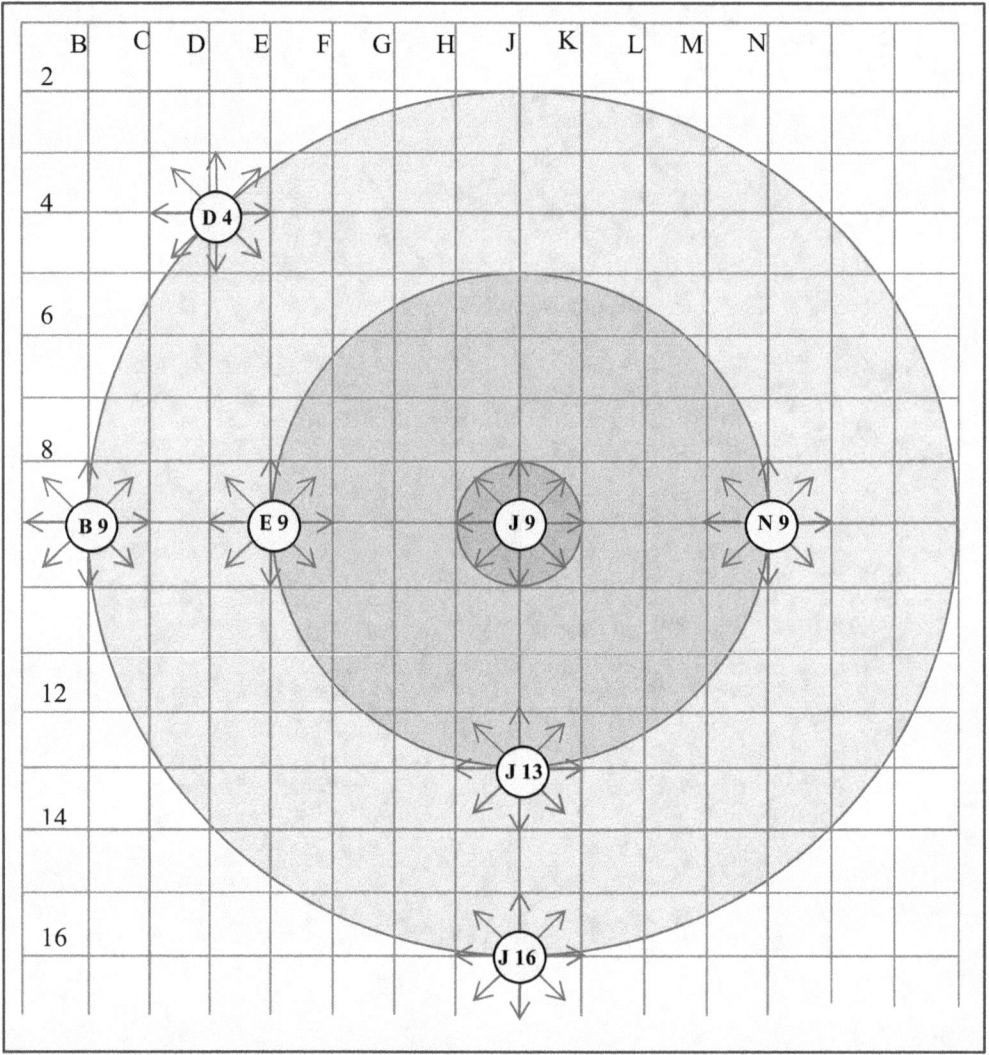

Not shown here, but a lot has happened in that seven seconds. Electrons, protons, neutrons, gluons, quarks, hydrogen, hydrogen molecules, and failed sub-particles, AKA (Also Known As) dark matter, have come into existence.

Receding Edge

Receding Edge II

The next demonstration is presented laterally with the picture on the right. It begins at the point of creation with light traveling to the right along the x axis from some point on the left. Time travels upward along the y axis. Let's pick a spot at the beginning and see what happens to both the receding edge and age of the universe we are watching.

Our starting point begins at the lower right corner. Beginning a few milliseconds after creation, we watch that location of our developing Universe. As time progresses, filaments, failed sub particles, gluons, leptons, quarks, and protons are manufactured as noted by the next level up. Meanwhile, new information comes in from our neighbors farther away implying the edge is receding by doubling the radius at row two.

We began our journey at a few milliseconds old, but since time is really unknown, we will now refer to neither time nor distance. Locations are relative positions only.

While plotting normally moves along the x axis from left to right, in this case we freeze a point and watch that point as it grows older. This forces the origin to move backwards from right to left. Since this is not a video, we must watch time grow along the vertical axis with static pictures. As time moves up the chart to row three, we see our mini universe has aged enough to generate hydrogen. We also see information arriving from new mini universes. As information continues to arrive from new sources that are farther away, it emulates a withdrawing process. This is why we say the edge is receding. This retreating edge reveals new matter in the form of filaments, leptons, and the makings of meaningful subparticles. In time, that new matter will also develop into the first element of hydrogen as our aged mini universe has done.

At row four, our location has aged enough to form molecules of hydrogen, row five gives us clouds of hydrogen, and row six failed particles have also formed into clouds. At the ripe old age of row seven, hydrogen forms into humongous groups millions of miles across. Failed subparticles have given rise to extra space, and the superforce has gathered large groups of hydrogen into ever growing super-duper formations.

Notice how the edge reveals various copies of the same startup particles at every age. As our original starting point has grown old enough to give us our first star in row eight, the superflux continues to bring information along the x axis from slightly younger formations of matter out to the very beginning of the universes.

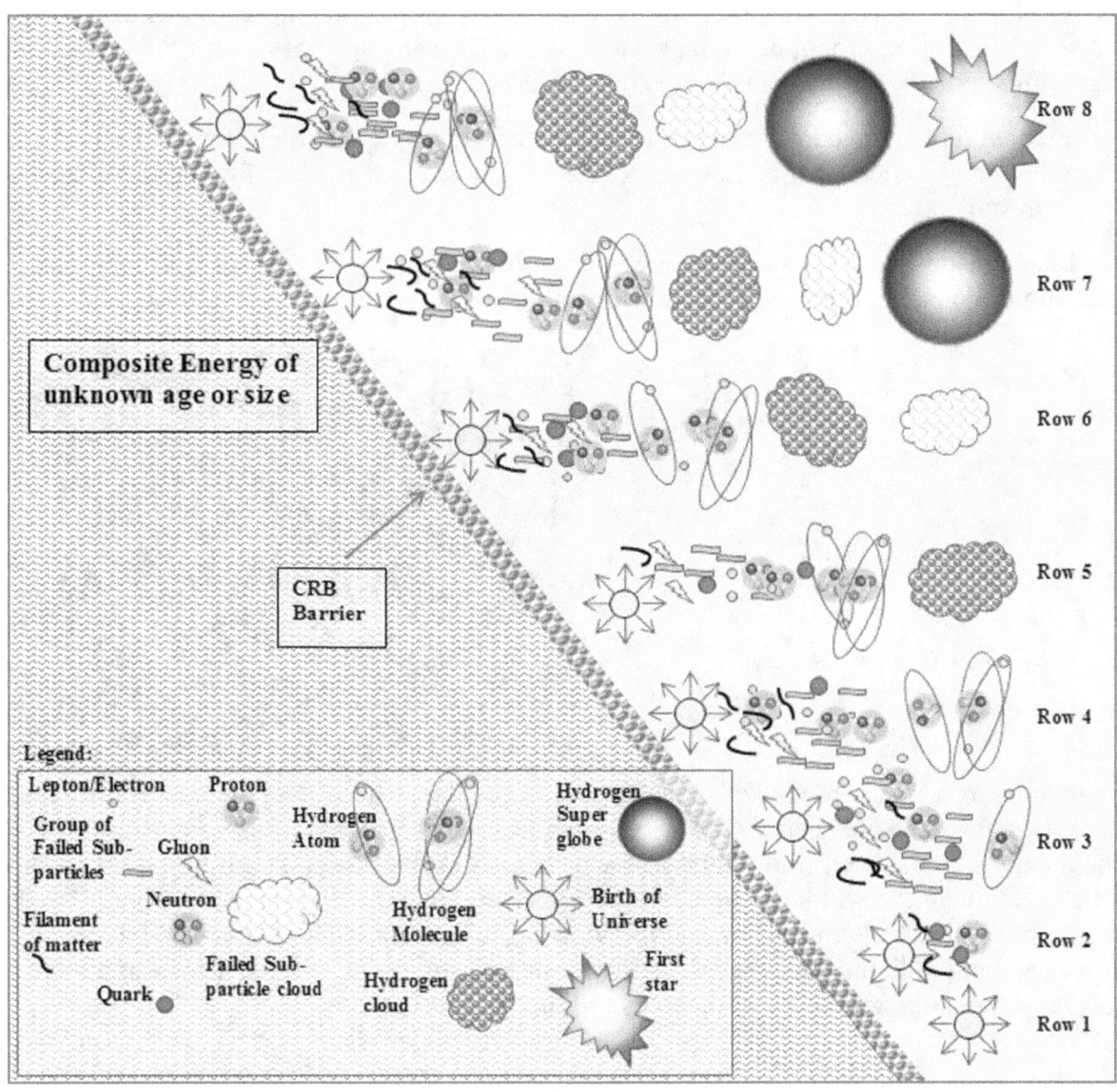

Receding Edge III

Maintaining the lateral format, we're going to jump to a point where our Universe is 7.1 BYO (billion years old.) This alternate view uses a WAG to come up with this age for the supernova that brings about our solar system. It could be any age before or after, but we are beginning there, there when our Universe is 7.1 BYO.

For the next three images, we want to concentrate only on one particular area. That area is where our solar system will be in the next 6.8 billion years. It is identified on the image as Pre-earth supernova.

The image below is a wide swath through our Universe at 7.1 BYO. It is a representation in three dimensional coordinates. The x axis, coming at you, represents one million light years

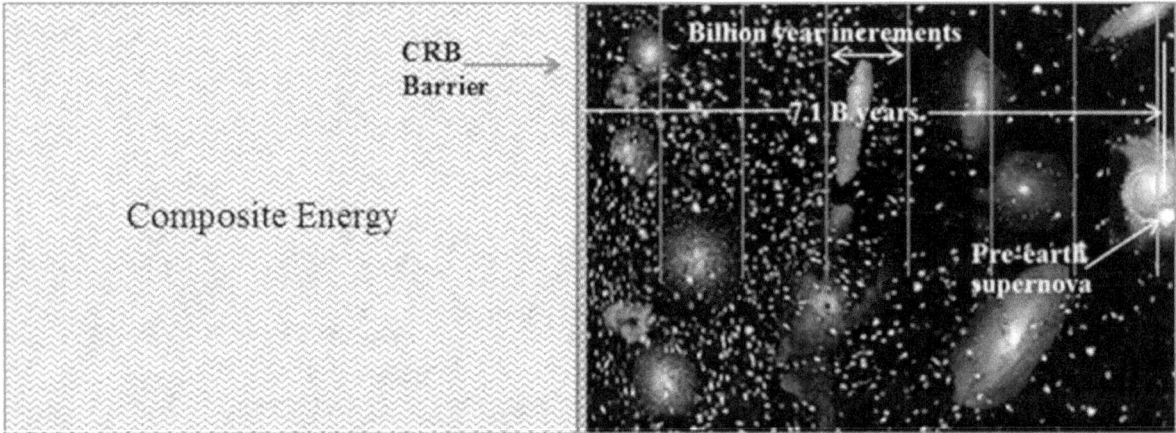

along the y axis. The left edge of the CRB represents the birth of the universe. Before that, there is only composite energy. The y axis runs from left to right, and the z axis is vertical. The z axis is that same million light year wide swath of the y axis. A million years on this scale is less than the width of the smallest visible line in the image, actually about 0.0005 inches, but it will have to do. Further, the view has been rotated 90 degrees and zoomed in so we see light radiating from the galaxies coming at you on the x axis. Those small white dots represent large groups of young stars billions of miles across.

Next, we look at the universe as it was at the birth of our solar system. It was a young 9.1 BYO with a full life ahead. Notice how everything near the birth boundary is still pumping out matter, but the large clusters of stars and quasars morph into large galaxies near the five

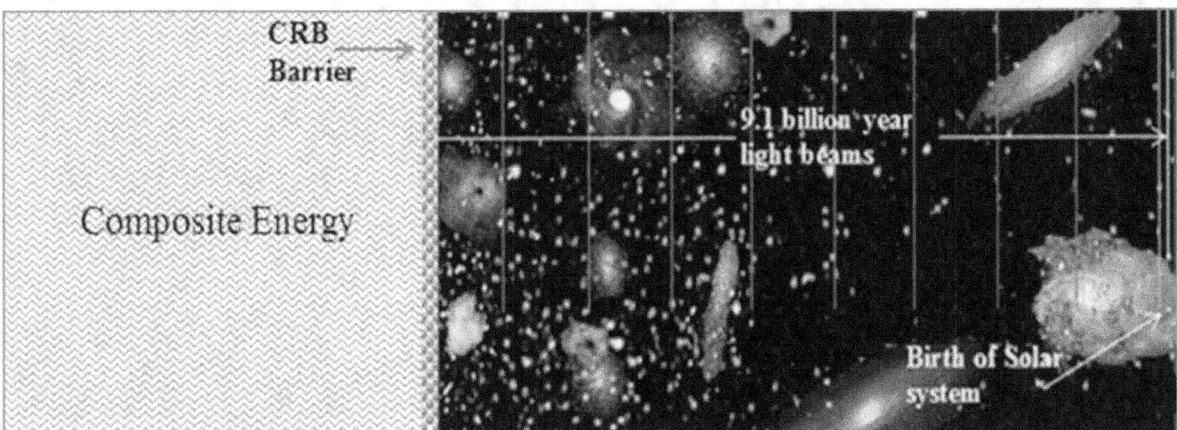

billion year mark. As failed subparticles create more black-holes, they gather star clusters and other wandering matter into galaxies of various configurations. It gives the appearance that space is growing, but it is not . . . (well, in a way it is), later for that. The movement of stars and groups of stars into galaxies emulates a thinning of space. The ratio of new space to new mater creation is approximately 5.25:1 at the CRB boundary. Then this new matter and failed subparticles will create even more galaxies. And that's the way it goes, on and on.

Next comes our beloved Universe. While a period of 13.9 billion years sounds like a very long time, it is not. When compared to its age and how much our Universe has left to live, it is only a drop in that old proverbial bucket.

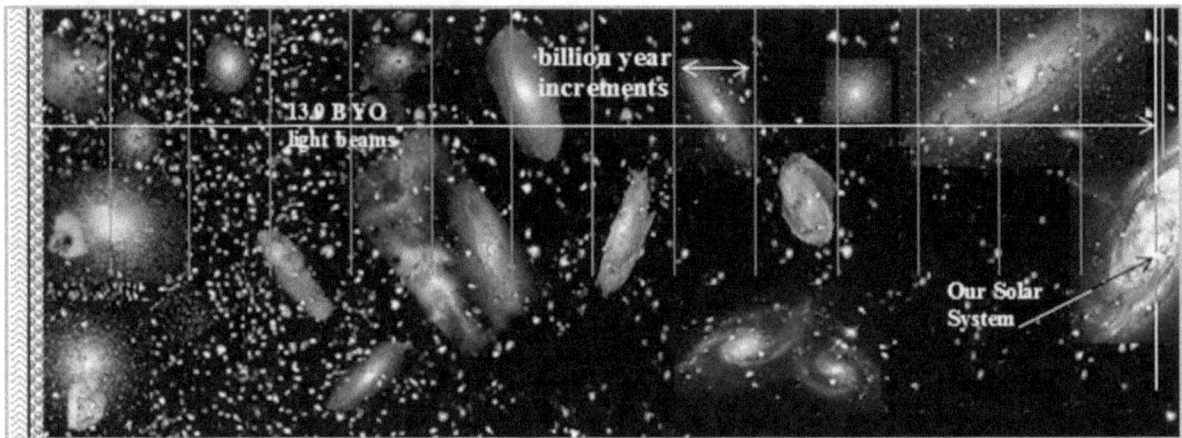

It appears as if the large groups of stars at the beginning have disappeared, but they have not. They've just moved from their birthing place to form up with millions of others around a black hole created by failed subparticles. Compare the 9.1 BYO image to the 13.9 BYO image and notice how Andromeda has moved towards the Milky Way little by little in the upper right hand corner.

As we observe the last three graphics, we can visualize how the universe expends into the area left vacant where space and matter is born. The universe just replaces composite energy with itself as it did from when it was 7.1 BYO to its current age of 13.9 BYO.

Chapter 5

Superforce At Work

The superforce gathers matter by way of transparency. There is a relationship between the number of flux lines that exit a bundle of matter to the number of lines that enter that same bundle. The ratio of outgoing to incoming is the transparency index. It is the opposite of opacity. If something is 100.00% opaque, it is 0.0000% transparent and vis-a-versa. If a body of matter is 0.0000% opaque, its transparency index is 1.0000. Indices of objects built of matter range from a transparency index of 0.0000 to 0.9999 with nine's going on forever. A filum is solid; therefore has an index of 0.0000. A black hole will probably have an index of 0.0000 or as close as it can come to it.

The superforce that did not convert to matter remained behind for a purpose. It held composite energy at bay such that space could exist between matter and super energy. Without that pressure, super energy would have closed in on the newly formed matter. Therefore, we can declare that space is comprised of the remaining superforce. Our Universe is pressurized. Space itself is pressurized. It applies pressure to every piece of matter because that tiny piece displaces the superforce, AKA, space. Pressurization causes expansion, and when our Universe reaches its maximum size, pressure will no longer exist and neither will gravity in any form.

There is further bad news to some: there can be no grand unification of fields if the superforce, known to others as gravity, did not separate as the other fields did.

The atmosphere applies pressure to every particle of matter. The ocean applies pressure to every particle as well. Buoyancy comes about because of a pressure differential. It applies to the atmosphere as well as water. A ball will rise to the surface of a swimming pool because a difference in pressure against the bottom of the ball versus the top forces it to do so. A hot air balloon rises toward the top of the atmosphere because the bottom has more pressure acting against it than the top. The superforce pushes two objects together because of differential pressure against opposing sides; in other words, the facing sides have less pressure than the opposing sides. The body's transparency index brings about this differential pressure.

The superforce consists of flux in the form of vectors. They come from one edge of the universe and continue across to the opposite edge. They come from so far away their source is almost flat. It is to a certain degree. Vectors are parallel within $2.6 * 10^{-15}$ degrees for a portion one light year across and parallel for an object the size of Jupiter's orbit. And for every incoming, there is an outgoing. Bi-directional, they are.

From the image below, we see that there is an equal pressure surrounding the orb, but the body is made up of smaller objects. Vectors are so small they go through an atom's nucleus with no problem at all. They continue until some smaller body down deep inside the larger object stops them. Many more make it all the way through and come out the other side unphased.

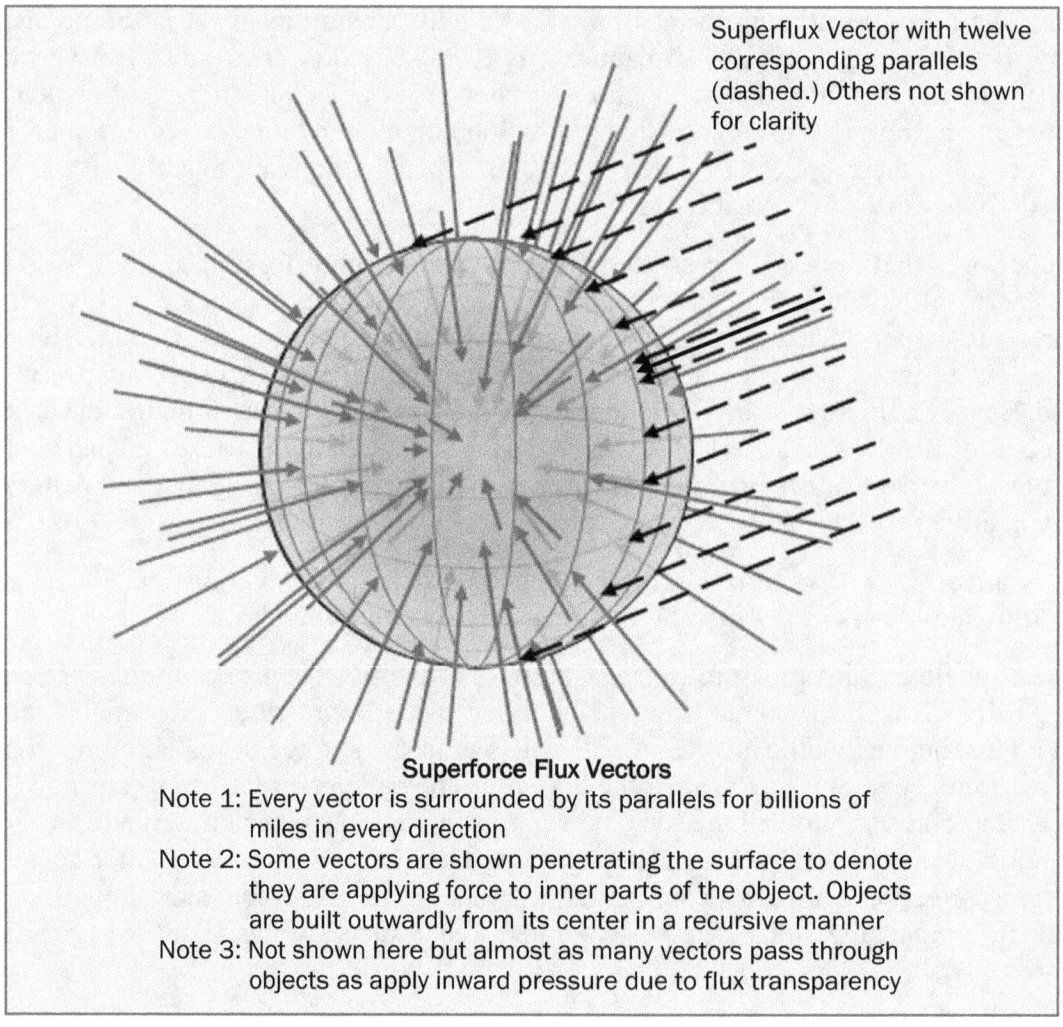

Superforce Flux Vectors
Note 1: Every vector is surrounded by its parallels for billions of miles in every direction
Note 2: Some vectors are shown penetrating the surface to denote they are applying force to inner parts of the object. Objects are built outwardly from its center in a recursive manner
Note 3: Not shown here but almost as many vectors pass through objects as apply inward pressure due to flux transparency

We can see that the globe cannot go anywhere because an equal number of flux vectors apply pressure on every square millimeter of the object. For every force pushing left, there is one pushing right. For every force pushing on the top, there is one pushing from the bottom. But what happens when something blocks the incoming vectors on one side? It moves—moves toward that less resistance.

Maturing Universe

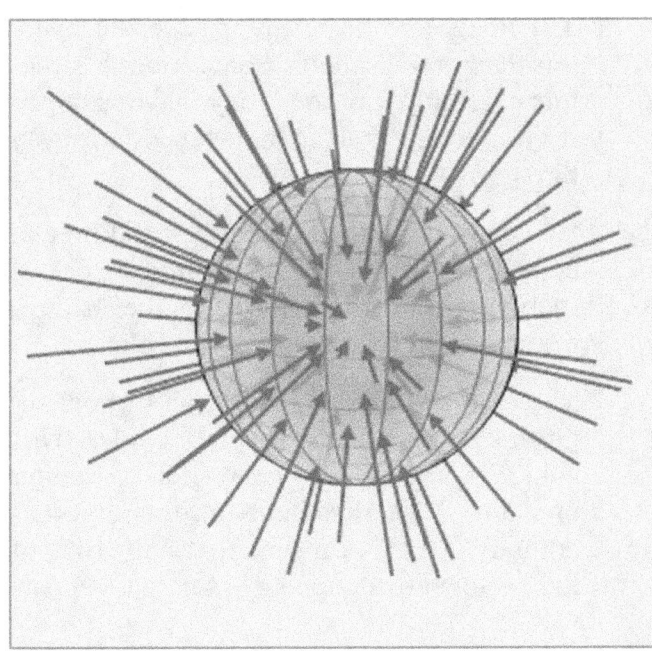

Let's take a smaller globe and experiment with it.

Using the object on the left, we'll place a magic hand blocking the incoming flux from one side.

Below, the dashed arrows and curves indicate the original position. By blocking some of the inbound flux on the right side, the globe has no choice but to move in the hand's direction. Of course the hand has the same force pushing it towards the globe, but since it's magic, it stays put. If one is not careful, a person may interpret that the hand attracted the object by way of an imagined power.

We can see from the demonstration that any difference in pressure causes movement. This is where transparency comes into play. A byproduct of transparency is a shadow.

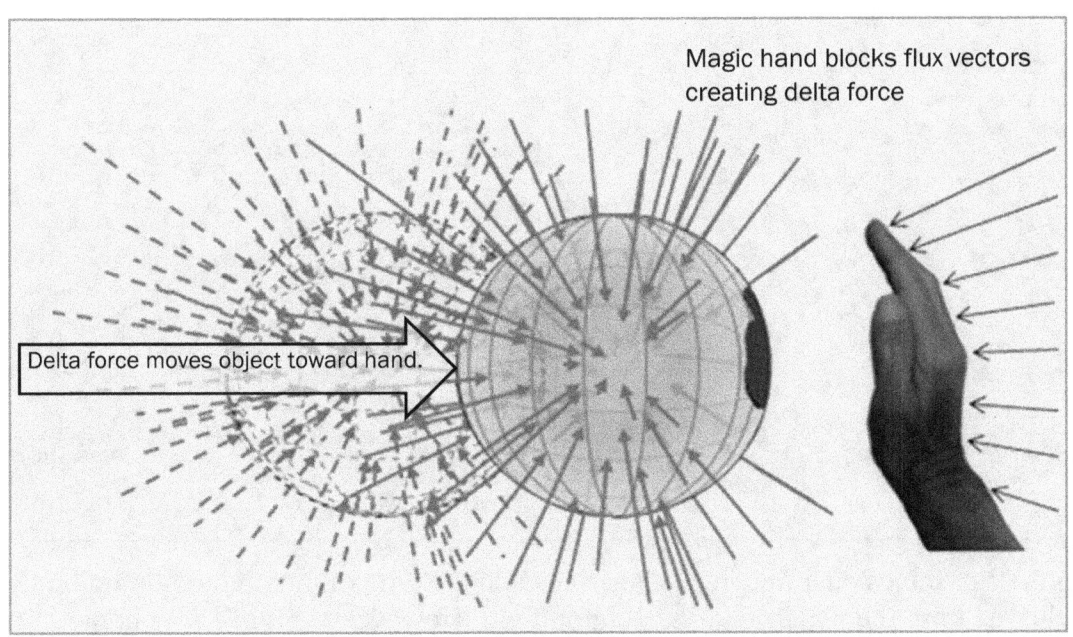

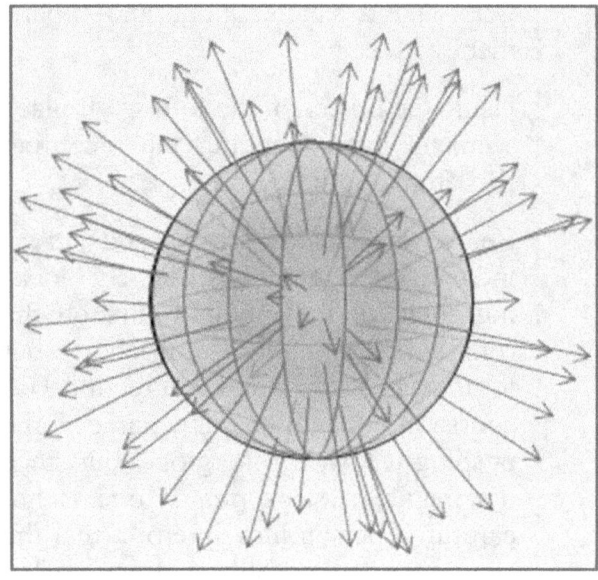

The image on the left represents flux vectors leaving a sphere. Some atomic structures making up the body have stopped those vectors that do not get all the way through.

However, looking at the globe head-on and comparing it to the incoming flux surrounding it, it is darker. That is the shadow, the shade left by the background flux.

The image below is a rendition of how superflux appears using special goggles. This could be any heavy planet such as Jupiter or Saturn. The flux index on this object is around 90%. That means that only 10% of the vectors are stopped by atomic level particles making up the planet.

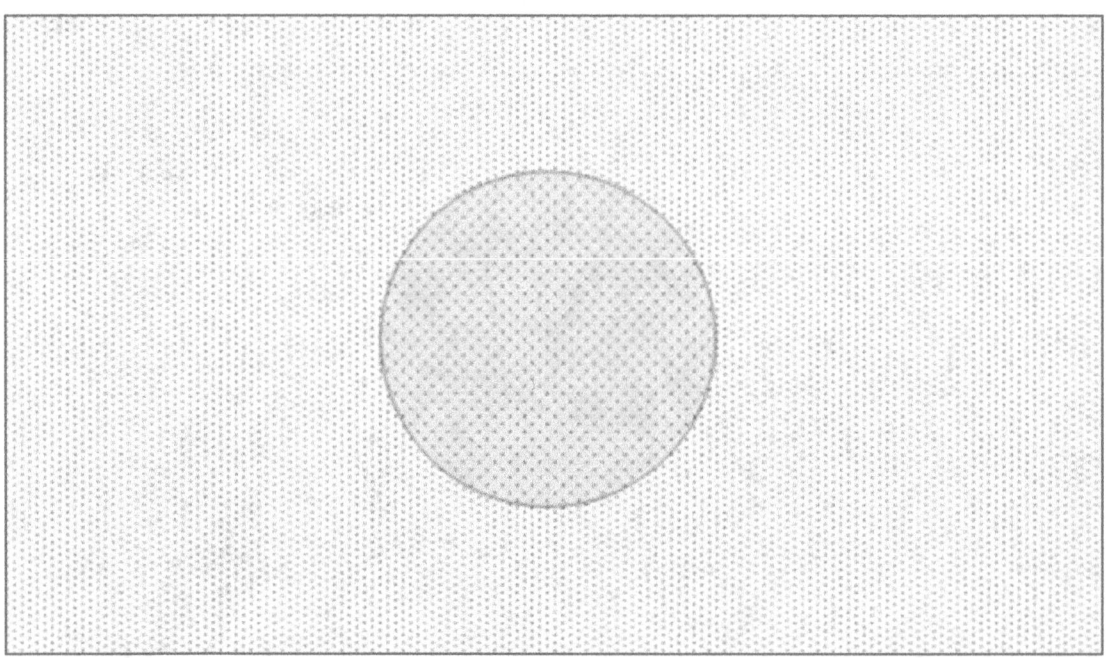

When the sun shines on a tree, it produces a shadow. Some of us used to rely on that shade to shelter us from the blazing sun in a cotton field. However, the sun's rays come from ole Sol himself while the shadow of an object, be it a planet or a sledge hammer, illuminated by the superforce comes from every place, from every angle. For that reason, the next example will have a single slice cut through space and the object itself. It is a plane slicing through the area.

Distant Objects: How objects find each other

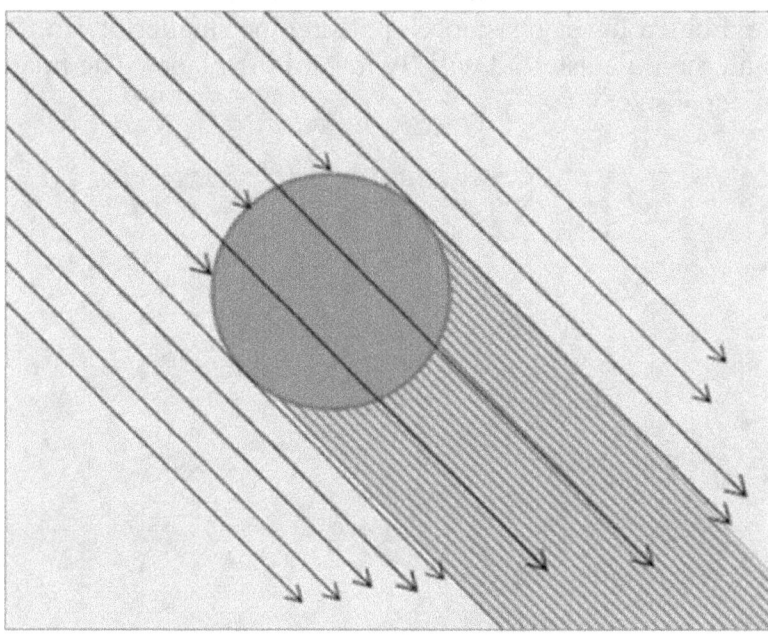

The object on the left is alone and billions of miles from its nearest neighbor. Once again, observing the image we see that the flux lines are parallel. And they come from every position of our Universe's outer edge. In other words, every square bit of surface matter has the superforce applied to it at various angles as well as every point inside. Since every proton, every neutron, and every electron making up the object displaces the superforce, each receives a vectored force. For the moment, we are interested in only one plane from one direction. We are looking at a slice of that plane including the slice of matter it dissects.

Again, we state all things are transparent to this flux. It can range from 00.0000% through 99.9999% transparent. Of course, 00.0000% transparency would be a black hole, and anything 100% would not be matter. It would be some form of energy. Atoms are 99.9 percent space so there is plenty of room for flux to get through without touching anything. If a neutrino can fly through the earth without touching anything, superflux has no problems sailing through as well. Even the earth is transparent to a very large degree.

Flux vectors are one step up from a line, and looking head-on, they are one step up from a point. If one zooms in on a flux vector, it will never become larger. A point has no diameter; a line has no thickness. A flux vector does have thickness that is just greater than a line.

This concept is important because it is what allows all objects to close in on one another. The flux transparency shadow between two objects produces less force leaving an unbalanced pressure tending toward the two. It is this unbalance that brings the objects together. In no way are they sucked or drawn together.

If another object wanders anywhere within this shadow, the vectors will be less dense; therefore, less pressure is applied to the visitor's surface. And since less force is applied on that surface, the visitor will move toward the object. For this purpose, the object is a small planet. In addition to the planet's shadow, the visitor will have one of its own. The two act in sync and move toward each other. If the planet's body has more material per cubic meter, its shadow will be darker providing a lower resistance opposing the smaller object. By deduction, the visitor is lighter and the little asteroid will fly toward the planet. The image below shows two such objects of equal size.

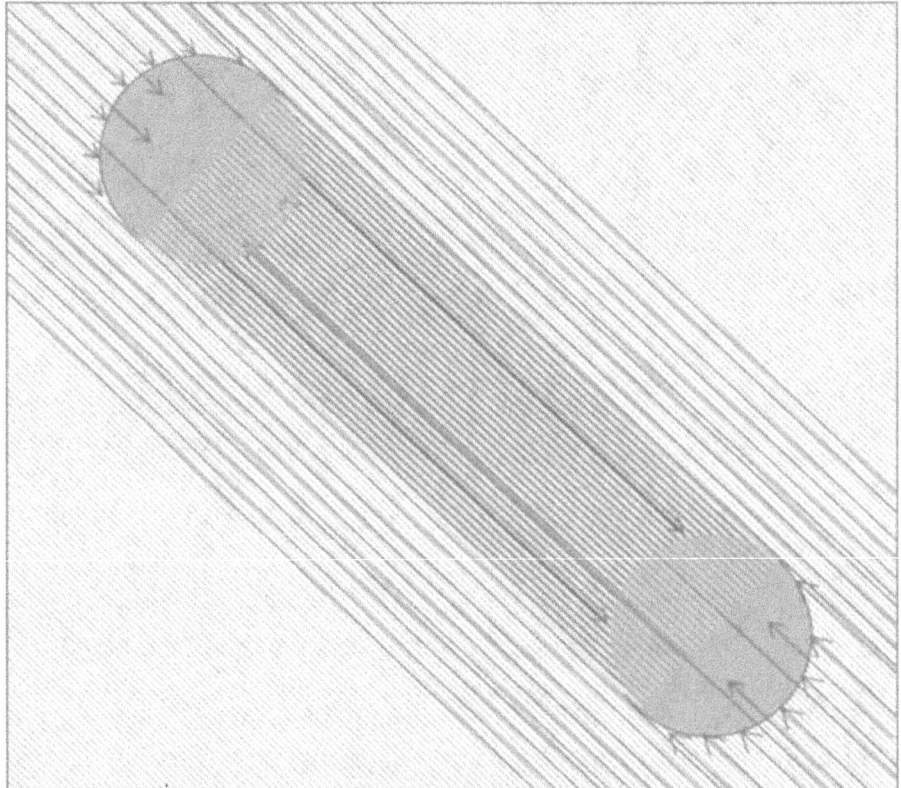

Now, it doesn't matter whether or not the object is a molecule of hydrogen or a stone of granite. The principle is the same. The superforce acts on an atom the same way it acts on a huge molecule. Building small molecules into ever larger ones is how the earth, sun, and moon were made. Not shown on the previous illustration is the more complicated interaction of more than two objects. They all combine to become a much larger unit. Then that unit combines with thousands of others and on and on *ad infinitum*.

As mentioned before, as long as an object is alone, billions of miles from its neighbors, all forces are equal which results in no movement without external influence. But when a small transparency shadow strikes another object, even as small as a pinpoint, there is always less force on the shadow's hemisphere resulting in movement toward the smaller pressure.

Warning: Without context or known history of the objects, it is impossible to determine whether they are circling one another or on a collision course. A single snapshot of the previous illustration cannot determine that, the more snapshots, the more accurate the information. The purest of all is a moving picture. The single snapshot above could be of two large objects orbiting each other at a high rate of speed. It doesn't matter how fast they are revolving, the shadow is always the same. The only difference is that when in orbit, a radical change in momentum is involved, but the differential forces are the same.

Understanding the flux shadow is of great importance; therefore, we are addressing it in several ways.

There's an old saying, "When you're a hammer, everything's a nail." Well the same goes for flux, "When you're flux, everything's a sieve." Flux sees the earth's crust as a sieve, the mantle as a sponge, and the core as super-hot steel wool. The superforce would probably describe the earth as a ball of steel wool wrapped in gauze wrapped in screen wire.

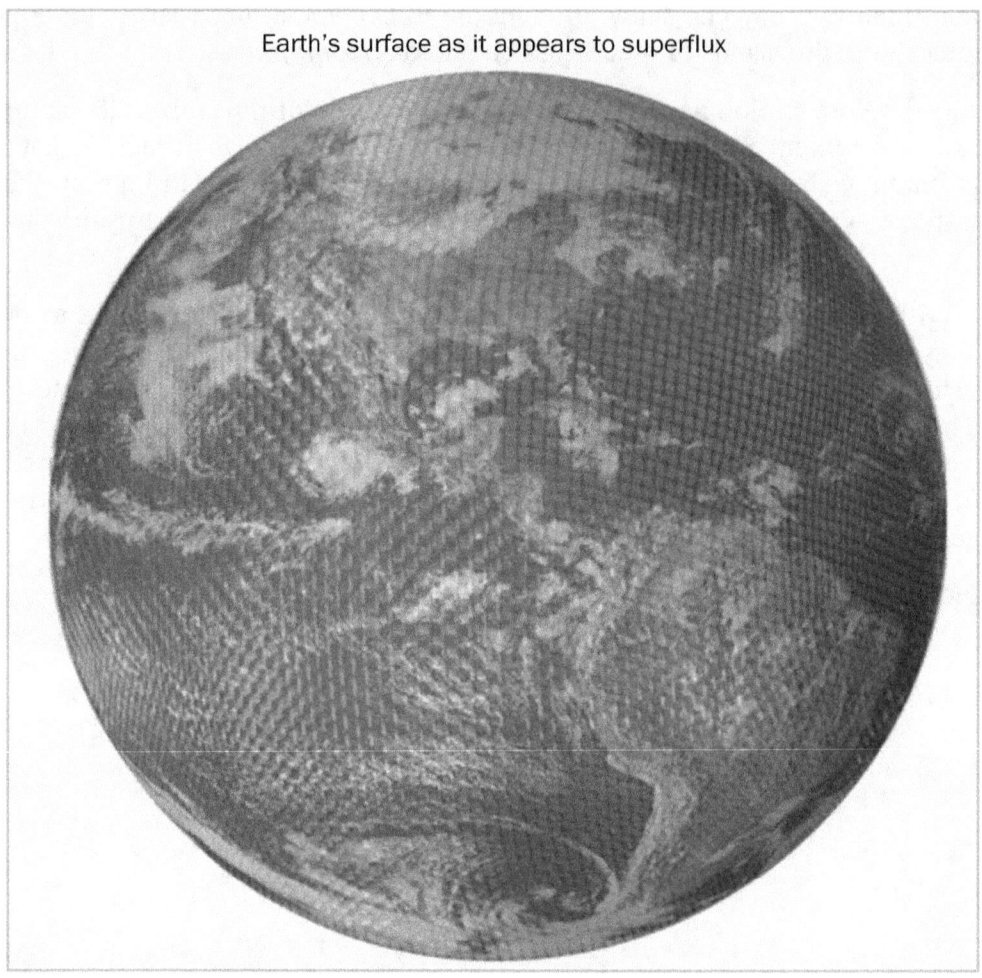

Earth's surface as it appears to superflux

Vectors that exit the earth's crust create a tidal effect when something comes between the earth and inbound flux from the distant edge of our Universe. Using a little imagination, one can see what happens to water molecules when a body blocks the vectors that are pushing them toward the center of the earth. Since there is less pressure driving the water towards the center, the outbound vectors have less force to overcome and lift that portion of water towards the blocking object, in this case, the moon.

The moon acts as the magic hand and causes all atomic structures to move in its direction. Rocks and other solid objects are subjected to the same forces but do not flow as easily as liquid, so off toward the moon H$_2$O goes. The solids move in that same direction, but that movement is only a few millimeters. The term solid refers to its everyday meaning because as far as flux goes, nothing is solid except filums and black holes.

Shadow Goes Global

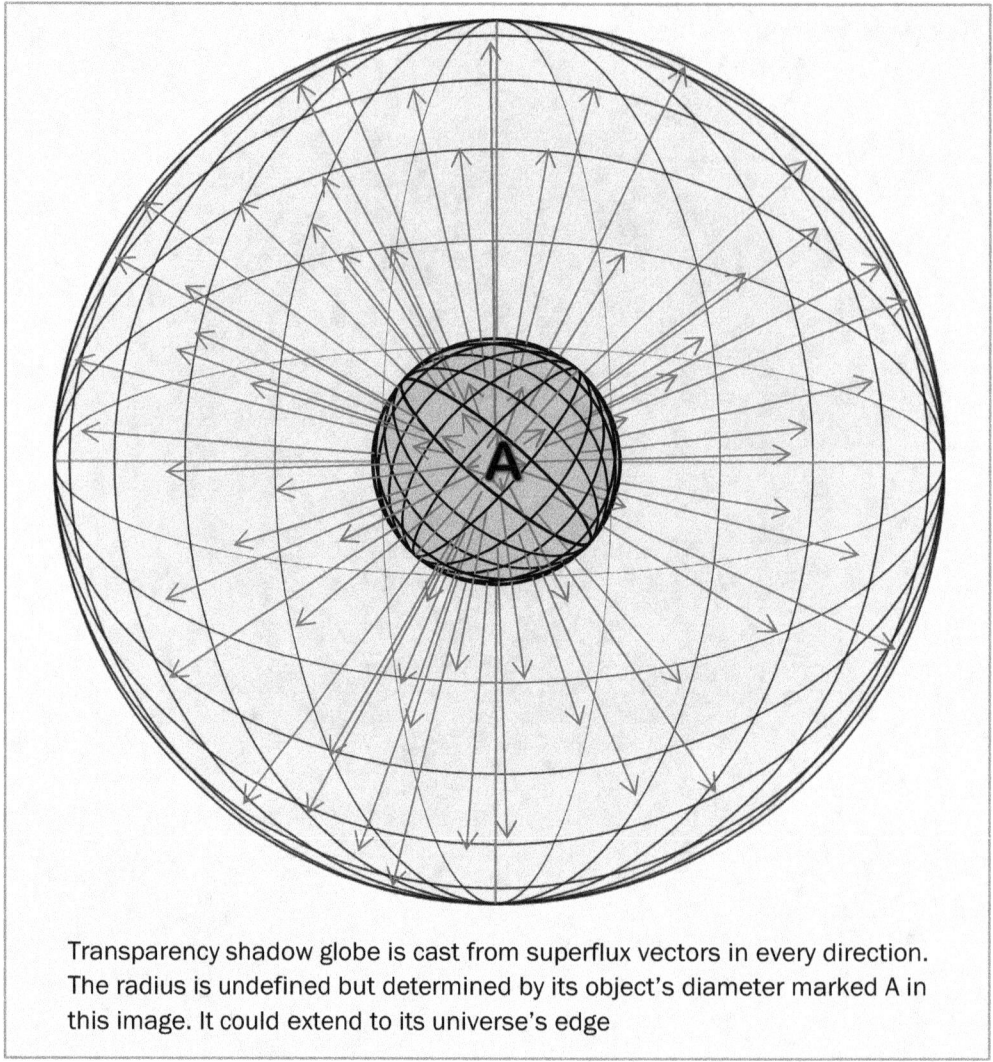

Transparency shadow globe is cast from superflux vectors in every direction. The radius is undefined but determined by its object's diameter marked A in this image. It could extend to its universe's edge

The image above represents a sphere in outer space. Even at a great distance, it will influence other objects, smaller or larger. It could be so far away it's only a dot in the sky, but that dot will affect other heavenly bodies.

Another picture of the global shadow and we're off to another subject.

The lower image shows how each shadow has another just like it at its 6 o'clock. The image identifies only two pair of the large and unidentifiable number. We could probably apply the term googolplex here, or better yet, any of the invented numbers greater than that.

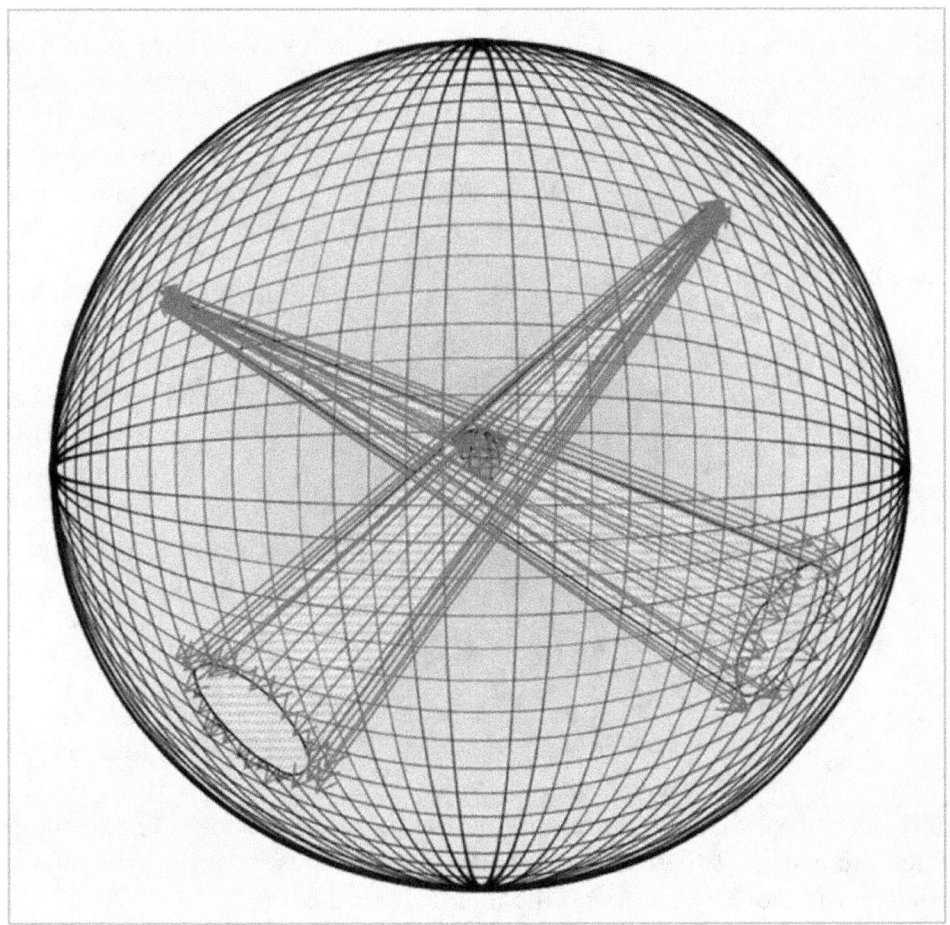

Chapter 6

Tides

Tide 1 represents a look at three separate flux shadows, and how they play a role in the moon/earth's tidal forces. The image is almost that of an eclipse represented by a penumbra/umbra diagram, but not quite.

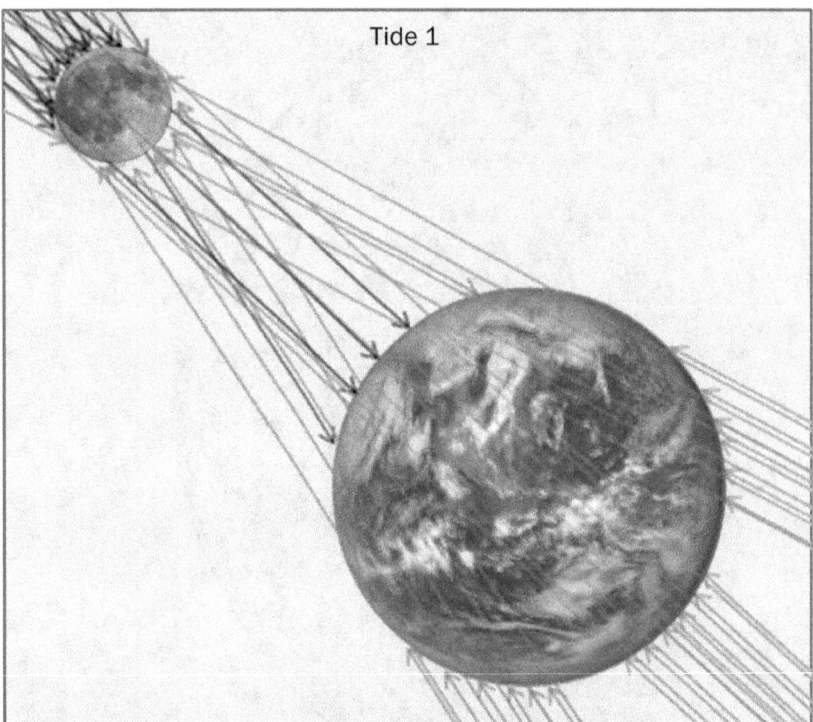

For this rationalization of how tides operate, we will observe these groups and their opposing vectors one at a time. The first thing to do is to concentrate on the center group. That is the group with the strongest influence: the group entering from the lower right corner.

However, the representation is too busy as is, so we'll strip the extra stuff away and look at a plane that cuts through this part of the universe including the earth. It is a slice through the center of the moon and earth that includes its solid iron core at the center surrounded by the molten outer core, the mantle, and the crust.

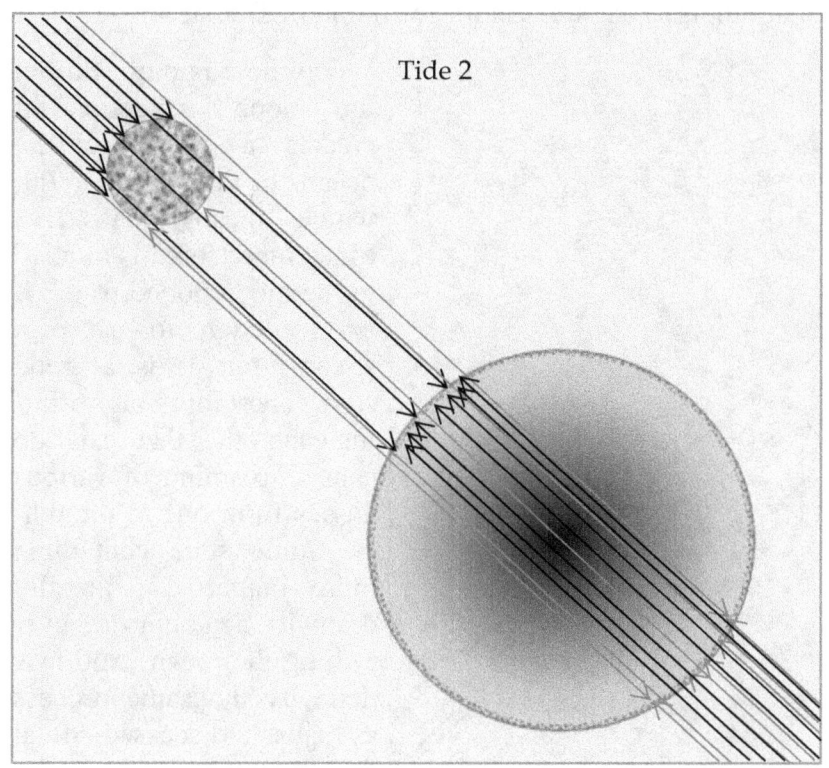

In this case, the three black arrows in Tide 2 going all the way through the moon represent that portion of flux that make it through unscathed and hit the water on the surface of the earth. They are in opposition of the seven dark arrows representing those that penetrate the earth and attempt to push the water toward the moon.

Flux that makes contact with molecules head-on in the direction of another object has the greater influence. All others compete by applying a vector additive approach that ranges from a difference of zero to slightly less than head-on force. A simple example of that process is in the appendix because it is important to understand how every flux vector, no matter the direction, affects the bundle of matter.

Next, we zoom in on water that attempts to leave Earth for the moon in Tide 3.

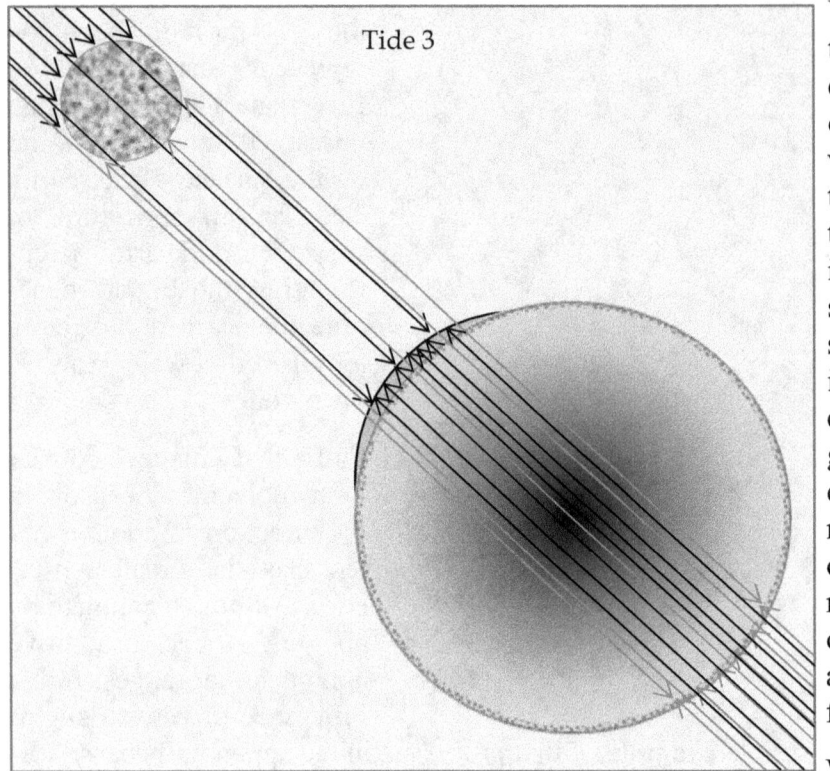

When there is more flux in the moon's direction, it creates a bulge in the ocean. In this case it's 7:3, which means that 2.33 times more force is applied to the moon bound water. It isn't enough to cause a stream, but it is a good start. There are stars orbiting each other that really do cause streaming of various gasses from one to the other. Some stars containing more matter can literally drain its neighbors of everything they own. And it is done by the same process as just discussed—tidal forces.

When one thinks about it, tides are just another view of how bodies are gathered via differential forces applied to clumps of matter. Every one of those clumps has a transparency shadow that brings them together.

The previous discussion is restricted to the group of vectors making contact to water molecules in line with the moon. Of course, flux comes in from all directions. The image in Tide 4 identifies the two groups coming from other angles.

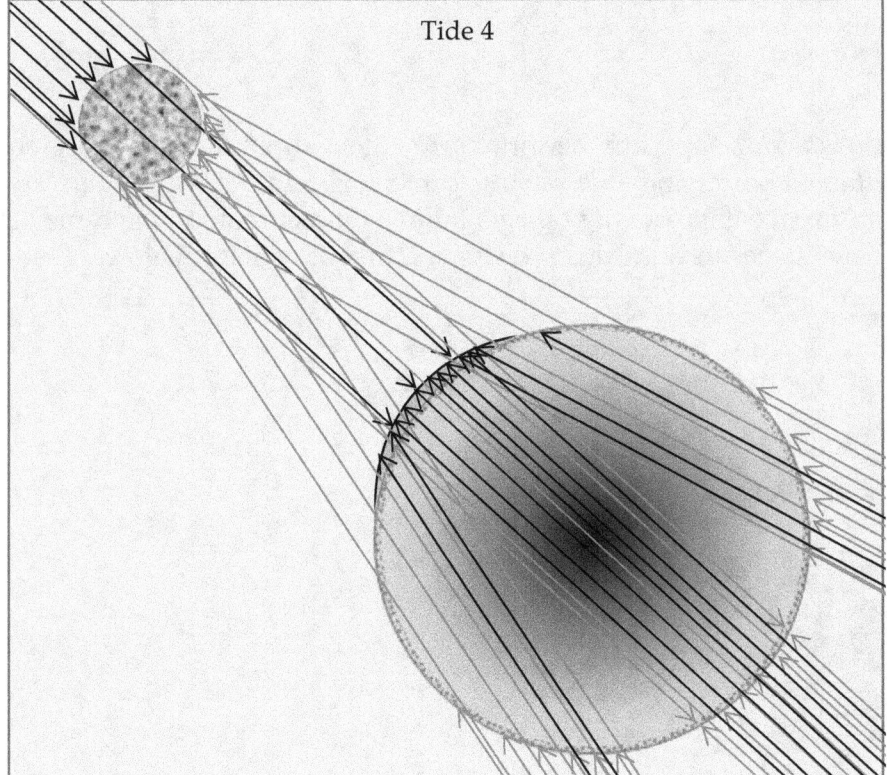

Tide 4

Chapter 7

Center of Transparency

There is another mysterious action superflux performs. How does it put the heaviest globs of matter down deep within an object and always at its center? The answer is that flux creates a funnel or pipe to insure the approaching objects follow a direct route to the center of the target. It will take a few images to show this process, the first of which is below.

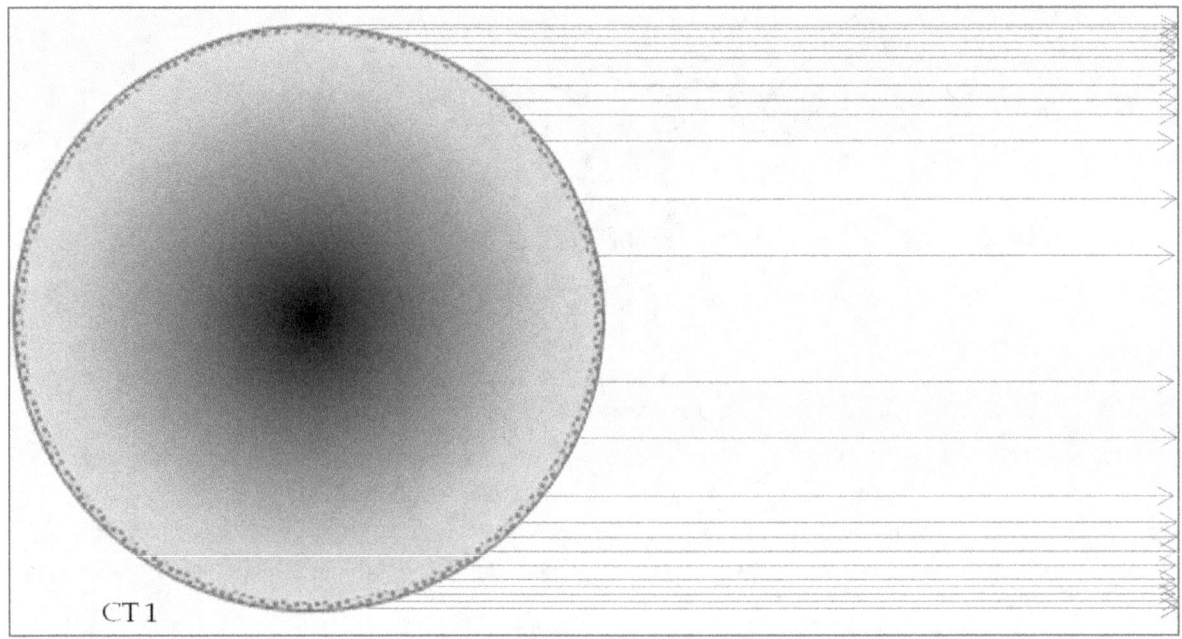

CT 1

CT 1 displays only the vectors that make it all the way through the earth. Inspecting the top and bottom of the image, we see the flux penetrating the crust. Notice how those flux lines are closer together than the ones at the center passing through the solid iron core and its molten outer layer. Two things affect the transparency at the crust: the less density of the crust, and the distance the flux travels through it. The distance flux travels through a globe is determined by a cord cutting the sphere, or in this case, the disk. The cord's length begins at zero, at 0°, and continues until it is equal to the diameter of the object.

An increase in the cord's length means more material will block the flux by way of increasing chances for each vector to contact some atomic structure. Fewer vectors make it through the middle of the earth than those coming tangential to the crust. That is, those that are horizontal to the outer layers are more likely to make it through. Since flux is thicker on those edges opposing anything inbound, an object must veer off towards the less resisting center of the earth.

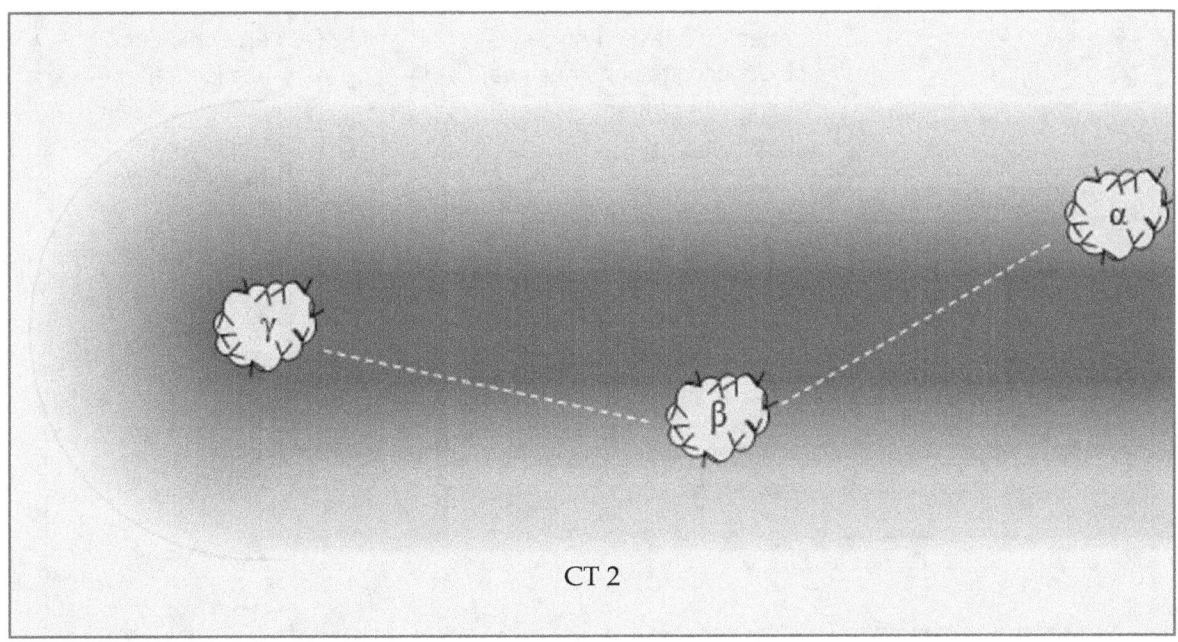

CT 2

Replacing our flux goggles with proper filters, we have a profile view as a body approaches the earth. Lighter areas mean more resisting flux while darker areas at the center mean less. It's like a funnel; an incoming object must follow the darker shadow.

Say the young earth is a half million years old. It is molten, thousands of degrees in temperature. Some estimate the early temperature between 6,000 and 8,000 degrees F. Anyway, it is liquid rock along with other ingredients. Then, here comes a megaton asteroid of iron. At position alfa, it meets a higher resistance that forces it to the dark side. It overshoots, and at position beta, the same resistance drives it back into the center of the channel. At gamma, it finally comes to rest in the young earth's center of matter. At that location, all forces are coming at it equally, and it remains there through today along with thousands of other iron meteorites. Other vectors reinforce the tunnel effect which will become apparent when we discuss bending light rays.

Bending Light Rays

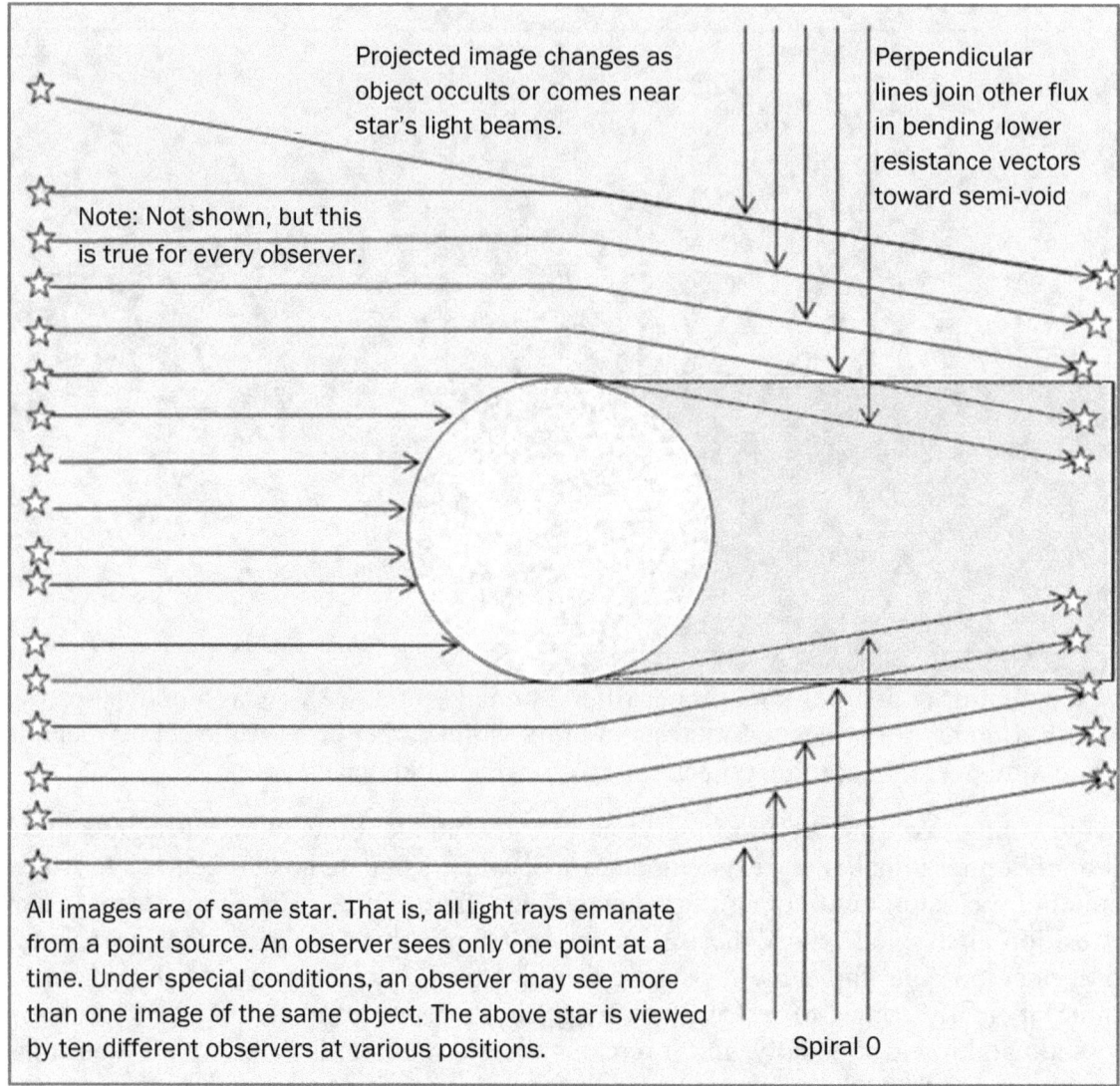

Traveling along a desert highway on a clear night is a terrific example of point source objects. The stars appear to travel along with you and the vehicle regardless of speed. Those same stars appear identical in a vehicle 100 miles up the road, ignoring Earth's curvature and terrain of course. What is really happening is that your eye is detecting different beams as you move along the road. You are passing through its superforce vectors the same way the object in the image above is passing through the single star's vectors.

The shadow is an absence of one or more vectors. Any time there is less of something, another thing takes its place. The density of perpendicular flux is greater than the density of the shadow allowing the greater force to rule. Only flux can affect flux, and the result is filling of the semi-void with redirection of vectors.

Of course, large objects such a galaxies are not single point sources. They are a group of point sources. Superforce vectors find their way through our Universe by the serpentine action brought on by transparency voids.

Since mass is an effect of transparency, one can see how easily it is to misunderstand that mass is the driving force behind this phenomenon. We need a correlation between transparency and mass.

Chapter 8

Darkening Shadow

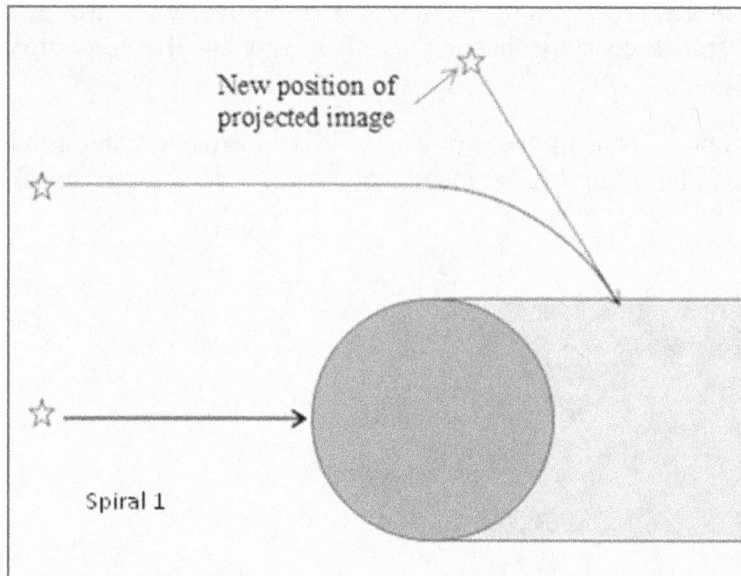

Let's watch two flux vectors as the object on the left becomes less transparent. That is, less flux makes its way through producing a darker shadow. The darker shadow influences the bending light beams even more which results in a longer arc. Actually, the arc is really the beginning of a spiral.

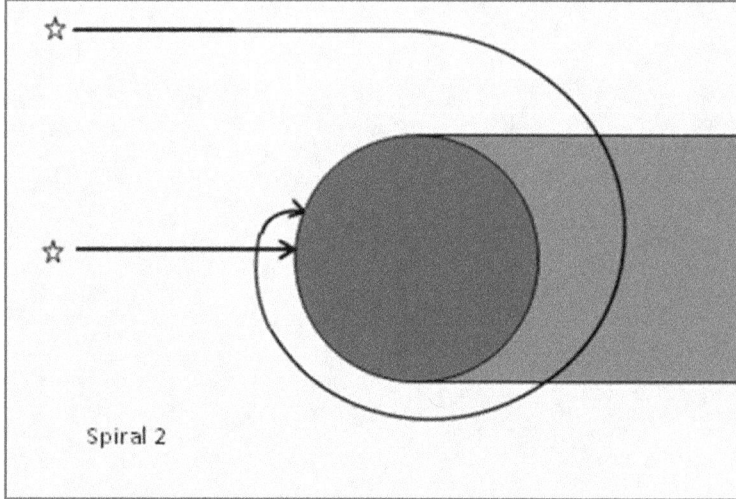

It's unknown where a runaway condition will occur, but at some critical point, the light beam's direction reverses, and the spiral continues to develop until it winds its way down to the object. The source of the light beam goes onto the surface, not around. Now instead of one beam there are two, and since light travels on flux, the force has doubled. The distance from the center of the object to the first spiral will become the inner event. Additional spirals quickly add even more vectors to the object's surface and internal structure once the runaway begins.

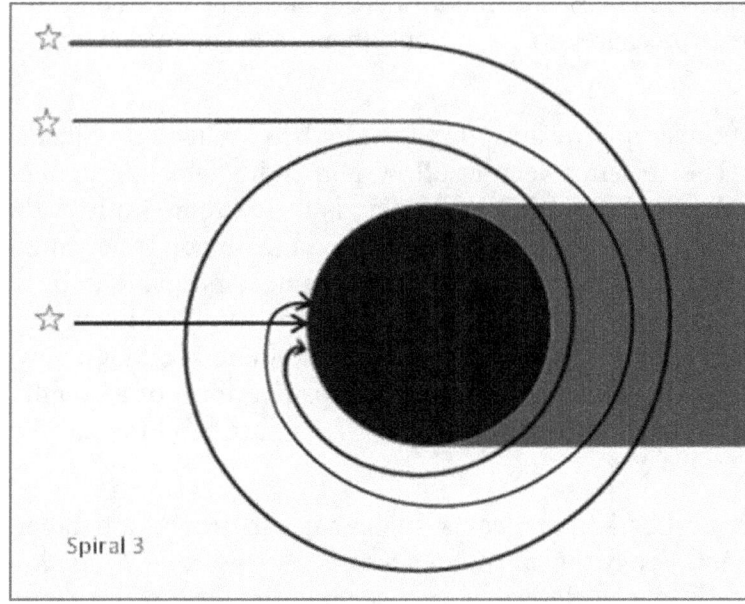

Amplification of the superforce begins in earnest as more outer layer spirals come into play. The exterior spirals become known as outer events.

The greater the radius of events, the greater the amplifying power of the flux. The zone is spherical so the force grows exponentially as a sphere's surface gets larger with an increase in its diameter. Thanks to the universe's power of two rule, every time this influencing radius doubles, the additional force brought to bear on the object quadruples. Another word for this inner-outer zone is the event horizon.

Now imagine what happens when an object becomes totally opaque. No flux can get through leaving a complete void of flux exiting the object. At this point all vectors are inbound. There is nothing to resist any object's approach.

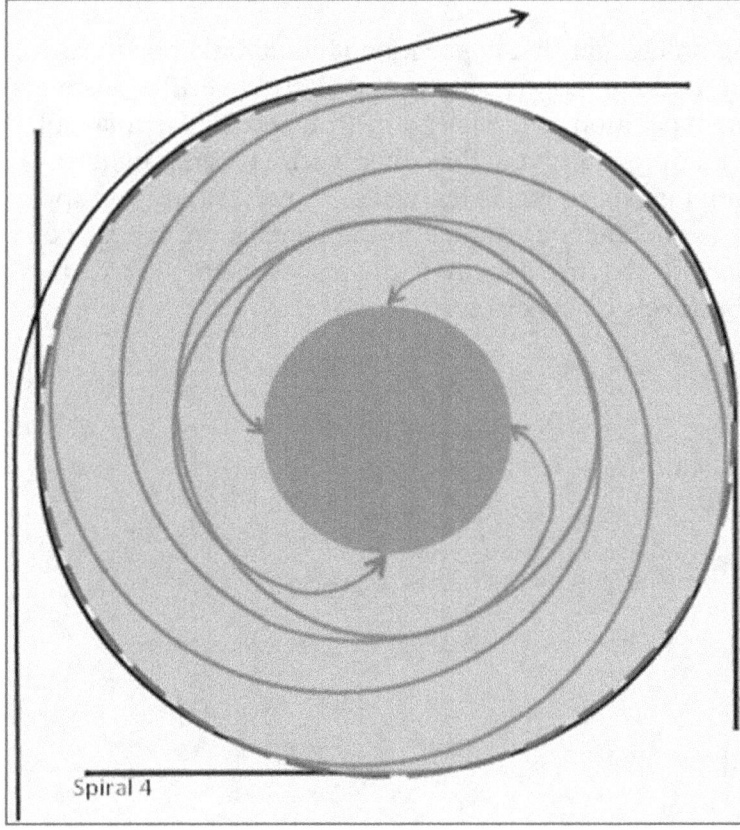

Spiral 4 represents a plane of four vectors, all swooping in adding their crushing contribution. Also shown is a single vector indicating that outside the event horizon, (dashed circle) they only bend to varying degrees depending on the distance from the boundary.

As the object becomes smaller and darker, the event horizon expands adding more force to the runaway condition. It's something like feedback entering a microphone. The ears continue protesting until someone takes correct measures to fix it. But, there is no countermeasure for a black hole.

See Spiral 0 on page 48 again for an example of how flux vectors bend when an object's transparency becomes too opaque. The missing vectors allow perpendicular and angular ones to bend the outer flux towards the void. Spiral 5, next page, is the extreme limit leading to additional forces exerting thousands of times more pressure on the object. It becomes a runaway condition leading to unknown endings. The event horizon does not have to be a globe. It may also be in the shape of an apple such that its poles are open to flux, incoming and outgoing. Some black holes eject streams of energy at its poles, so there is either a way in/out at the poles, or the energy escapes outside the horizon, perhaps in a form of a doughnut or torus. Outside the event horizon, the vectors do not wrap. They just bend to varying degrees.

A large galaxy or other massive units bend light beams such that supernova and other bright objects can be seen behind them. See Monica Young's article *Astronomers Predict a Supernova* on *Sky and Telescope's* web-site. It describes five separate images of the same supernova behind an elliptical galaxy.

Although light bending is an ongoing process, at first conception it could only be witnessed under certain conditions. Sir Arthur Eddington was among the first to verify Einstein's theory of relativity when he setup an expedition to the island of Principe for the total solar eclipse of May 29, 1919. This experiment proved that flux vectors bend light rays. However, scientists insist on declaring that space bends the waves. Since gravity and space are linked, it is understandable how astronomers can come to the conclusion that space is what curves light instead of the superforce having its way with an absence of flux vectors. Also, not everyone is aware that light travels on superforce flux.

Maturing Universe

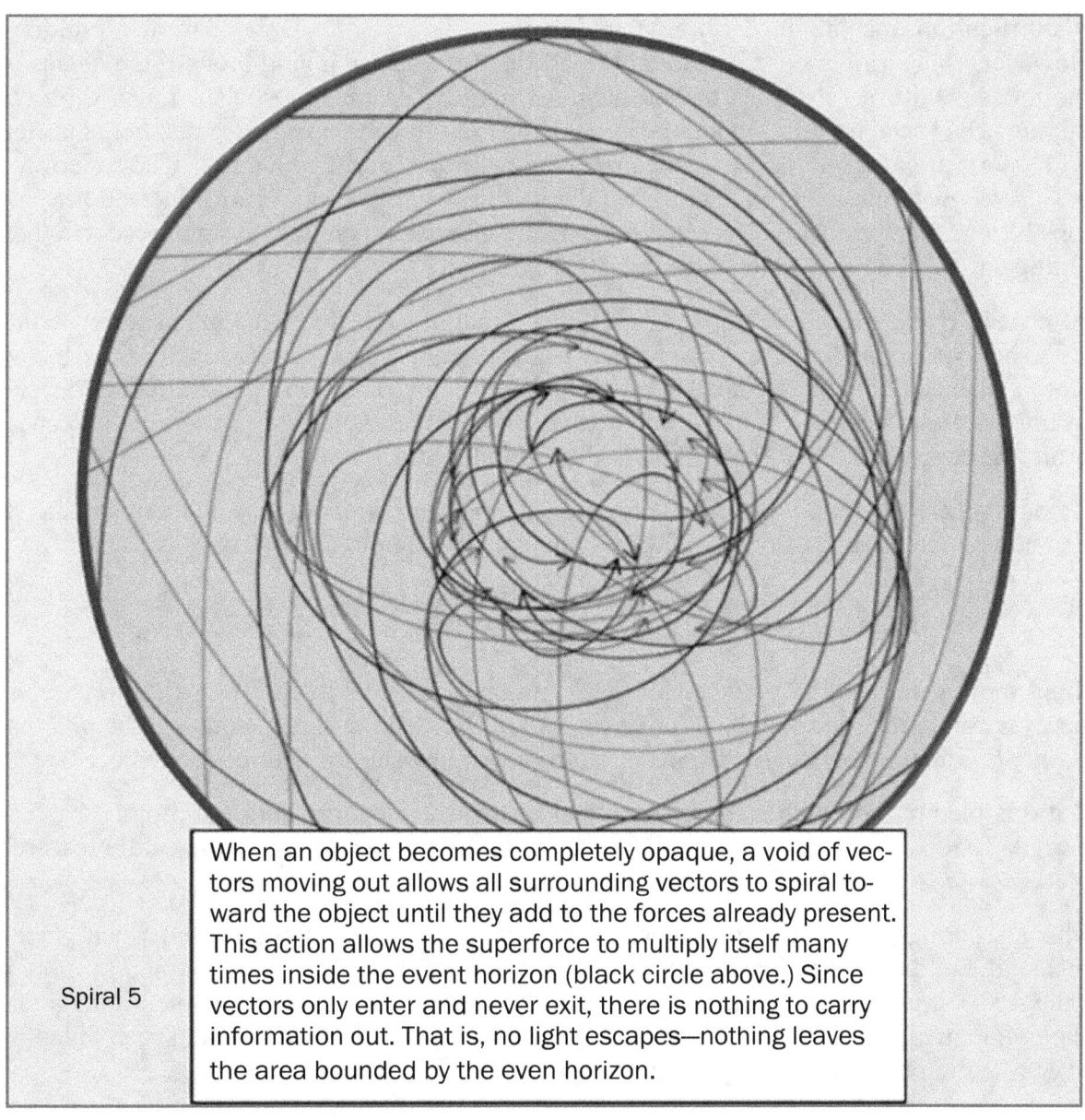

Spiral 5

When an object becomes completely opaque, a void of vectors moving out allows all surrounding vectors to spiral toward the object until they add to the forces already present. This action allows the superforce to multiply itself many times inside the event horizon (black circle above.) Since vectors only enter and never exit, there is nothing to carry information out. That is, no light escapes—nothing leaves the area bounded by the even horizon.

Neutron Stars

It's difficult to imagine the earth crunched down to the size of a large apartment building. However, comparing the size of an atom with its electrons swinging around its nucleus to the nucleus with its electrons tucked neatly inside the protons presents a much different picture. The space used to accommodate electrons in orbit disappears. Remember, a proton with its electron tucked inside is a neutron, and neutrons love togetherness. The superforce has its way with these guys and crunches them with very little opposition. Once all the protons become neutrons, it makes sense how nature can pack them inside small places. When letting air out of bubble wrap it shrinks rather quickly.

After nature removes all the room taken up by an atom's electrons, it's much easier to imagine how millions of neutrons can replace a single atoms space. It varies quite a bit, but an atom's diameter could be 10,000 times its nucleus' diameter. Comparing the size of the sun to this nucleus means the outer limits of the solar system would lie just inside Neptune's orbit: the edge would be 2,161,880,000 miles from the sun.

Using a sphere as an example even takes the example further. If we take an atom's radius to its nucleus' radius ratio of only 5,000:1 to calculate its volume, it would be,

$$\text{volume of sphere} = \frac{4\pi r^3}{3} = \frac{12.5663 * 5000^3}{3} = 523{,}598{,}775{,}598$$

times more space to accommodate all that extra matter. That's just over ½ trillion neutrons for every atom that once held that position, and over ½ trillion times more weight, and that is only one atom. How many trillions of atoms are in that star for neutrons to replace?

But this picture comes after the process that placed the electrons into the protons has occurred. That story is interesting, and you can learn how it takes place with a little research on the www.

This also brings on a question. Although an electron's charge is neutralized when placed inside a neutron, does the law commanding it to release magnetic energy when in motion change? That is a very important question. If the electron is still required to emit magnetic flux while in motion, that is a whole lot of flux lying in wait in such a small space. Imagine over ½ trillion times more magnetic flux just for one atom's worth of volume.

Do we need to rethink neutron stars and magnetars?

When an object spins, it tends to flatten, and the faster the spin, the flatter the object. As one might suspect, an object the size of Earth being squashed down to $1/1000^{th}$ its size would spin 1,000 times faster at its surface using only momentum as a reference. Of course if a volume of material is considered, it becomes more complicated because that volume contains more material as it gets squashed.

Planets do not morph into neutron stars, but we'll use the earth as an example for size reference. Say the earth rotates one thousand miles per hour (mph) at the equator, and it is 8000 miles in diameter. That means it rotates 1,000,000 mph when it shrinks down to eight miles wide. That is 11.57 revolutions per second. At this state, it is not a planet. It is not even matter as we know it because there are no electrons, no protons: just plain 'ole neutrons. A bunch of neutrons spinning round and round so fast each one needs to fly off in the current direction of travel. That direction is tangential to its rotation. That ever present superforce counters the centrifugal force attempting to outcast each piece of matter, so the outer neutrons can only go outwardly so far.

But what is happening to those near the poles? They are moving much slower and do not have the outbound force to contend with. However, they do have a slight differential force applied to the outside hemisphere, so they migrate also.

There is always a greater pressure difference at the poles of a fast spinning object than at its equator. As this differential grows, a slight flatting of the poles occurs. The faster the object spins, the thinner the once spherical's poles become until the superforce punches through and the object takes on a doughnut shape. It is a torus. Superflux passes through the poles and carries all the electromagnetic information in two highly focused directions. Such is the operation of a neutron star and a black hole of a Quasar.

A gyro's poles form two cones as it undergoes precession when a torque is applied. In the case of a humongous black hole, there isn't enough matter around to apply much of a torque. A Quasar may be in the process of pointing its laser in another direction, but earthlings won't detect the movement for generations of astronomers. But if a neutron star's neighbors can apply enough torque to induce wobbling poles, every time the beam passes Earth it will generate a signal. The timing of the signal depends on how the cone's ellipse points in our direction and the speed of the beam around that ellipse. If the angle of precession is 90 degrees, the signal hits Earth twice per revolution because there are two poles.

Maturing Universe

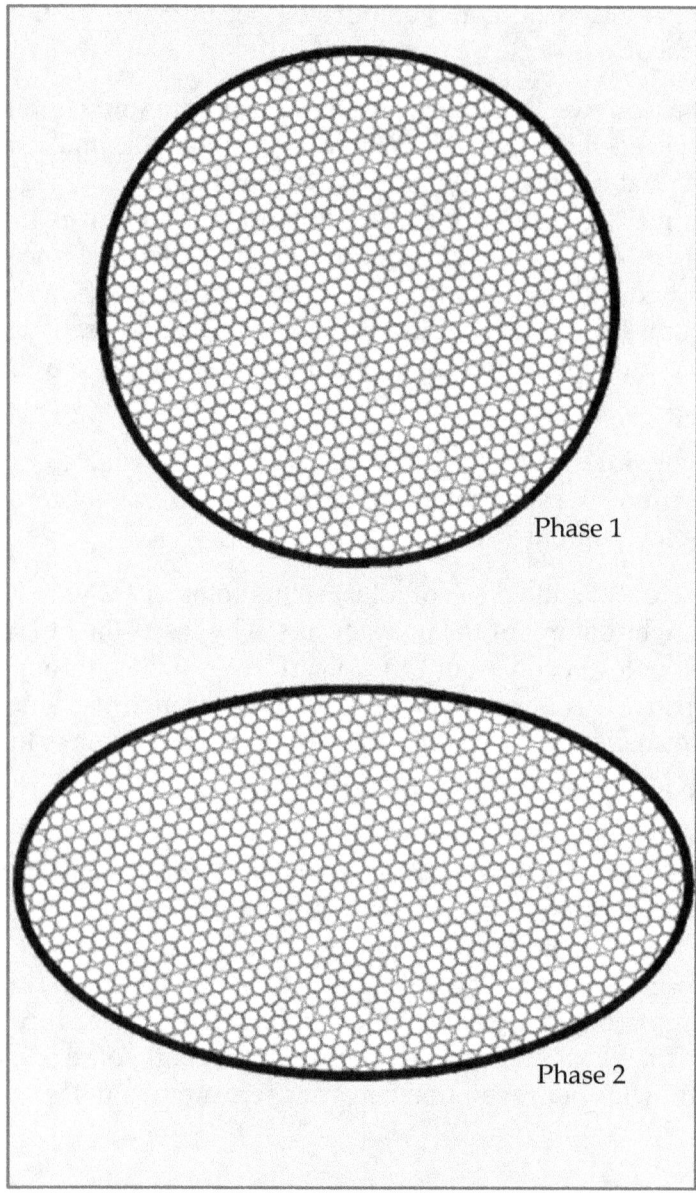

This experiment requires the use of ball bearings to emulate neutrons. The image on the left represents a small object filled with such. It is a sphere with little or no spin.

What happens when the object spins extremely fast—so fast that its matter spreads away from its axis of rotation?

Images Phases 1 thru 5 show the five possible phases as a neutron star flattens and potentially ends up as a torus.

At Phase 1, the object has not begun to spin at a high enough rate to contract, so the change is in pause waiting for shrinkage to begin in earnest. It only begins to flatten at a certain rotational speed. However, sometimes after the supernova starts to crush the once massive star down to a smaller object, it must increase its rotation rate at some point.

Somewhere near ten to twelve miles wide, the object has gained enough speed to send the balls outwardly, and as it shrinks even more, it must spin faster. Phase2 shows how the ball bearings have migrated outwardly while the superforce has flattened the poles.

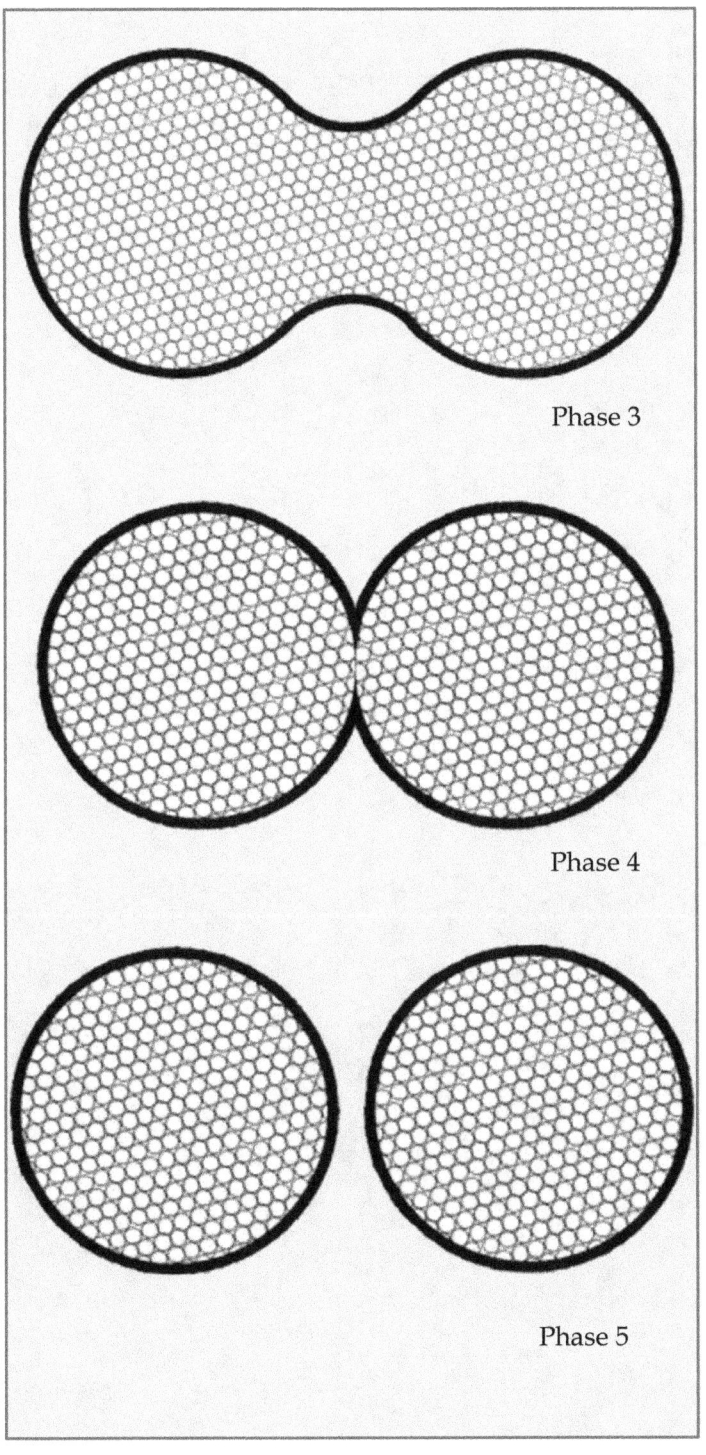

Phase 3

Phase 4

Phase 5

At phase 3 the bearings continue to migrate away from the center leaving less resistance to oppose any forces at the poles.

At phase 4 the superforce is ready to push through the poles and carry information in both directions. However, the greatest release of magnetic energy probably happens just as the torus moves from being a horned torus to a ring torus at phase 5.

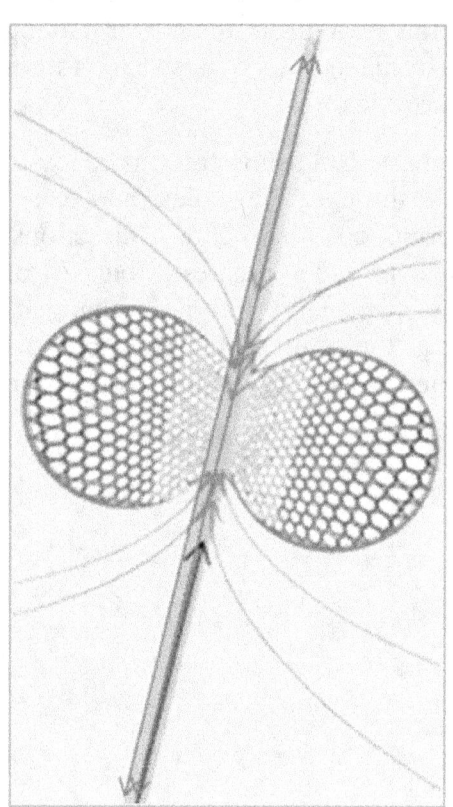

The image on the left is a neutron star that has spun itself to a point where the superforce has broken through but the torus has not opened.

Chapter 9

Powers of two in Operation

For a moment, we will step into the conventional world and speak of the superforce in its normal sense. Gravity.

Science students learn the gravitational formula for how two objects attract one another. Despising that word "attract" used with regards to gravity, we will suspend that spite temporarily.

For those unfamiliar with, or those having faded memories of, the formula appears below.

$$F = \frac{GMm}{r^2}$$

where F is force in Newtons, G is the universal gravitational constant, M is mass of object 1, m is mass of object 2, and r is the radius or distance between the objects' centers of mass.

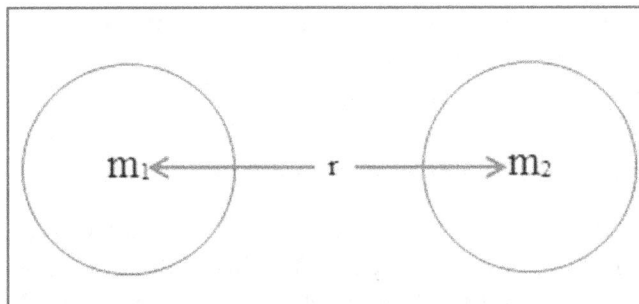

The two orbs on the left is a representation of the formula with only big 'G' missing. It is there, but it is unseen and all around and internal to the orbs.

The objective of this topic is not to review the formula but to describe why the force varies inversely as the square of the distance between. Why is it the square of the distance instead of 1.810 times the distance or 5.815 times the distance? Or for that matter, why it isn't any one of the other infinite number of choices? It is because our Universe demands rationality and stability. Earlier on, we mentioned the universe adopting the number two as a stabilizing quantity. It reflects a binary choice, yes-no, on-off, one-zero, but never maybe. Maybes turn into failed particles. Maybes turn into dark matter.

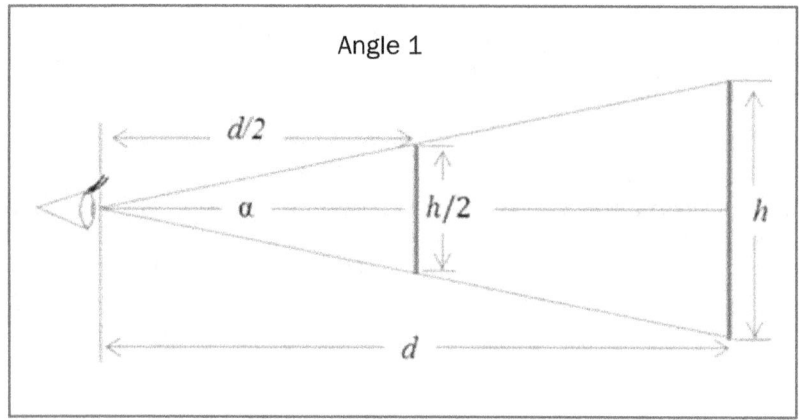

Using the similarity rule for triangles where all angles remain the same, the ratio of its sides remain the same. See Angle 1. Here a small triangle is superimposed on a larger similar triangle. At distance d, height h is fifty meters. At distance $d/2$, the height $h/2$ has been cut in half to twenty-five meters.

The same is true for heavenly bodies.

This requires another picture, an imaginary view of a distance object, even a pinpoint works. The imagination part is visualizing the superforce vectors coming directly at you from a great distance. They come at you surrounding the object and through the object in question.

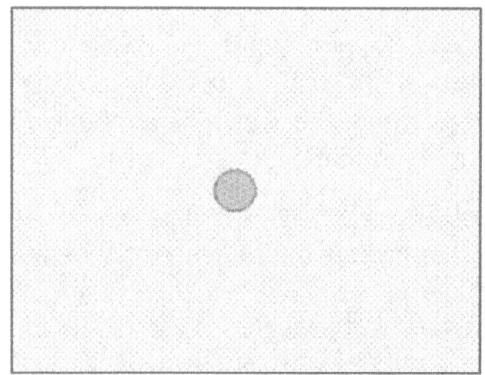

The perspectives of Angle 1 must be reversed for this to work because the object needs to grow as it approaches the observer instead of contracting as the distance increases. That is, the physical size of the object remains the same, but its arc increases. It takes up more of the sky as it comes closer.

There are two representations of our distant object. One on the left as it appears to flux detecting goggles, the other a profile as in Angle 2. Say it is an asteroid, a large one, perhaps large enough to be round. A radar observer first spots it at some point in the sky when it's just a blip on her screen. That point is z shown on Angle 2. The astronomer researches the blip and discovers there is no history. It's new; its orbit unknown.

A few days later, the new discovery has moved from its initial point 'z' to 'h' in Angle 2. It is also a known distance from Earth. Bold verticals depict the side view of the asteroid's disc, and line y-z is looking down the x-axis. The lines depicting the object's size is to scale relative to their height.

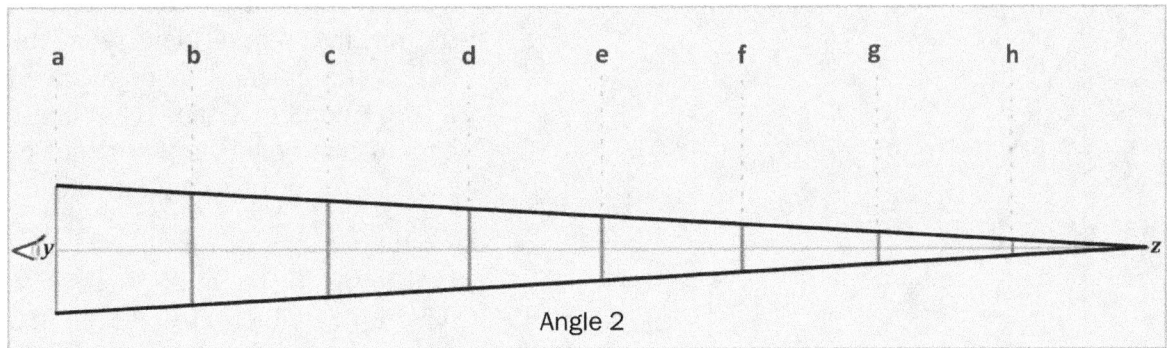

Angle 2

Objects in the sky are measured in arcseconds, but the chart has transformed those arcseconds to millimeters. And even those aren't exact because of page size, and on and on. However, the images are still close enough to grasp the idea of how powers of two influence the superforce over distance.

The astronomer has a good idea of the asteroid's size at position 'h.' A few days later, she can determine that the size has doubled at 'g' and its distance traveled since 'h' has also doubled. According to similarity rules for triangles, at position 'e' its height has doubled again since discovery. Finally at 'a' the proportions have repeated. The positions at a, e, g, and h are all places where the objects size in the sky has increased by a factor of two. Positions under b, c, d, and f will be used to plot interim values where the distance did not double. We chose fractions of the distance from y to z for ease of plotting a graph.

That is certainly a long way to go in demonstrating how a disc's radius grows exponentially as it approaches an observer. But it is also an important concept because as the radius of an object grows, so does its area. That is the key to understanding how effects of the superforce grow exponentially as two objects close in on each other. Since Area = πr^2, when the radius doubles the area quadruples. We must be careful here because r in this explanation stands for radius of image and the other r for big G refers to distance between objects.

Another graph is required to show how the distance between two objects affects the strength of the superforce acting on each body, or gravity to standard beliefs.

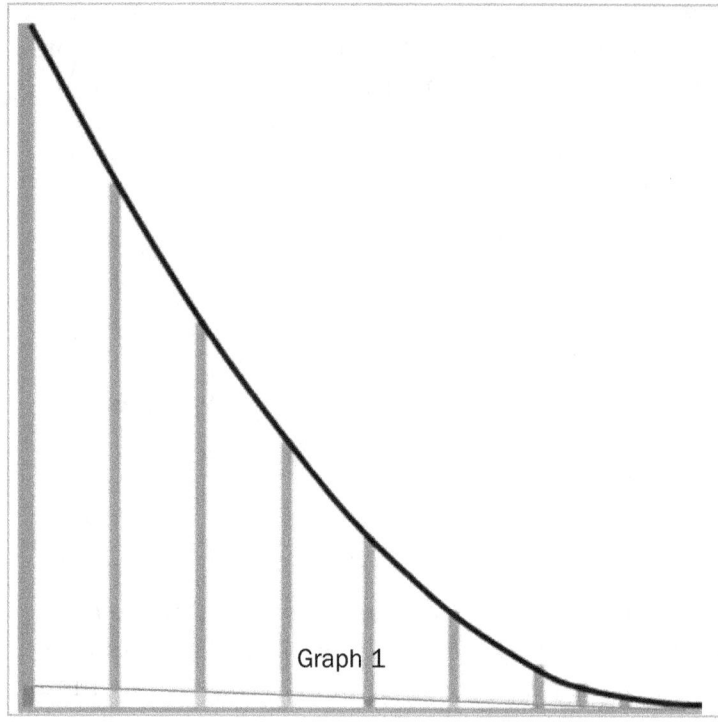

Graph 1

When the disc's area is stacked on top of its radius it gives a better view of how much the area grows with a slight change in diameter. Vertical lines have been placed on the original marks identifying the radius at a given distance of travel in the image of Graph 1. When a curved line connects the height of each value, it forms a parabola.

As the surface of an object quadruples, its differential force also quadruples which also means that the oncoming force has increased by a factor of four.

From the graphs, we can visualize how an approaching object blocks out more and more of the background flux causing the shadow to increase by the same proportion. That is, the shadow becomes darker exponentially. This increase in turn is what produces a differential force that has quadrupled.

The above example implies that the differential force is directly proportional to the square of the radius of the approaching object. This is only a part of the alternative formula; other factors must include transparency index of both objects and other yet-to-be-determined properties.

The intent is to visualize how the area of an approaching object leads to the force increasing as distance decreases and to demonstrate *why* the superforce, or gravity, operates on bodies as it does, not what it does. The WHAT has been known for hundreds of years. Understanding how superflux transparency affects the force should do that. From the preceding, we can infer that there is a correlation between gravity, mass, and superforce transparency. Those details are somewhere in the future.

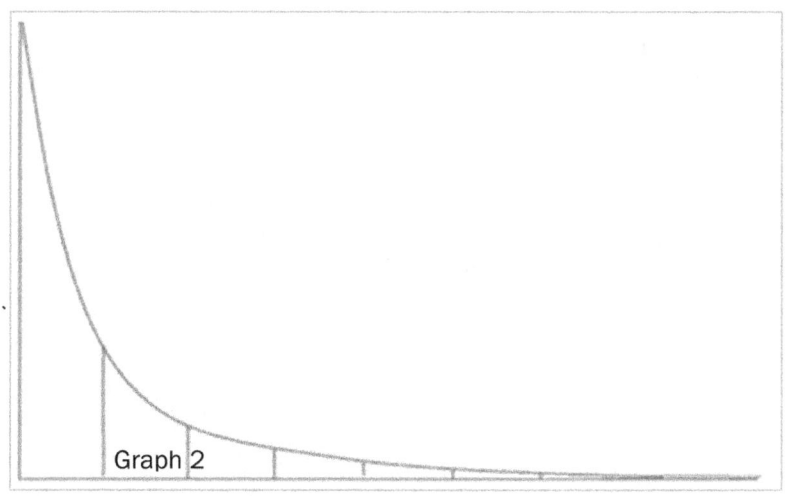

Graph 2 is another curve depicting how the inverse of the closure of distance affects the radius coincides with the previous drawing. In both cases the radius of the incoming disc is dependent on the distance between, one being directly proportional to the radius; the other inversely proportional to distance between. Notice how the value on the denominator increases very rapidly as the distance closes toward contact while the direct version is relatively slower.

For ease of plotting the inverse graph, the numerator (Gm_1m_2) has been evaluated to 1,000,000, and the denominator of r^2 takes on values of x^2 because the plot is along the x axis. This looks like a parabola drawn by the curve
$$y = \frac{1000000}{x^2}.$$

However, when using this equation the value of x can never get near zero. Neither can r in Newton's equation of $F = \frac{GMm}{r^2}$.

It is very upsetting when someone speaks of a person floating in a small room at the center of the earth. They reason that gravity surrounding that point cancels, and there is no more force. If the universe operated in that manner, it would probably be true. But it does not. The superforce crushes anything at Earth's center. The total weight of any column of material in any direction will come to bare on every particle of matter in the center. And that center will continue to grow smaller and smaller, perhaps even down to a point. There will never be an empty space at that location because the superforce does not attract, it crushes.

Chapter 10

Doppler Effect

The picture below is a bird's eye view of a train whistle while a steam locomotive sets idle at a boarding platform. It is a plan view of a slice through the acoustic sphere while sound radiates from the steam whistle as air is compressed and decompressed. Dark circles indicate compression while lighter circles show decompression, and somewhere between is ambient air pressure. The engineer has signaled that he is ready to pull away from the platform.

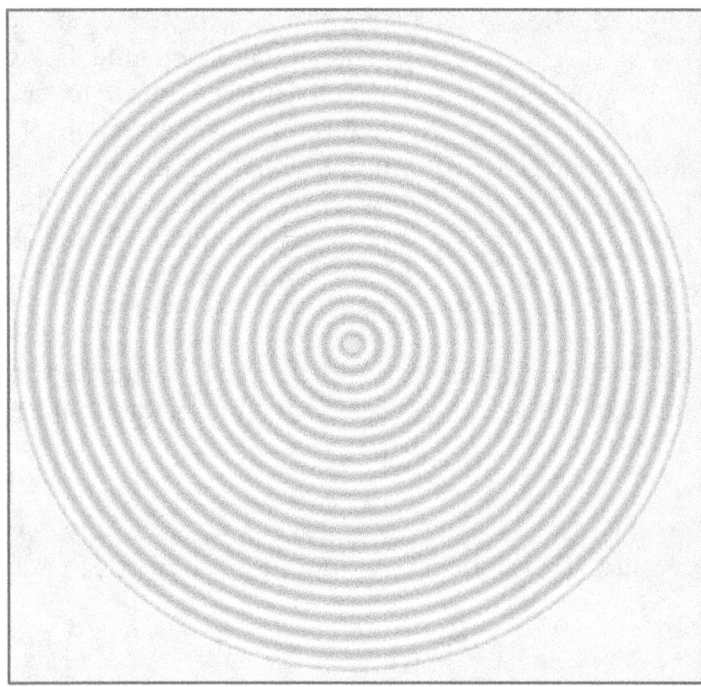

A few years before Christian Doppler published his work on the changing color of binary stars, people along the Ohio and Mississippi rivers recognized a slight changing tone a whistle made as a steamboat came at them then moved away. Later people along railroads would hear the same changes but at a quicker pace. Trains, even in their infancy, moved faster than any other transportation means. Unknown to the observers at the time, that change in frequency would become known as the Doppler Effect.

The frequency of light and sound changes with respect to an observer's movement. It seems there are various explanations for these phenomena, some of which are confusing. We hope to clear up some of the muddle. While chasing rational for the redshift of binary stars is a daring proposition, we will begin at a much slower pace. Since sound covers less distance in the same time period, our discussion of the Doppler Effect will began there.

Sound, radio, and light wave lengths are measured in meters. So instead of mph, we measure the speed of a train in kilometers per hour. For those interested, multiply miles per hour by 1.609344 to convert mph to kph. That is, 60.000000 mph is equal to 96.560640 kph. The significance must be great because later the time periods will be in billionths of a second or nanoseconds. Doing this up front saves a lot of ink and messy reading.

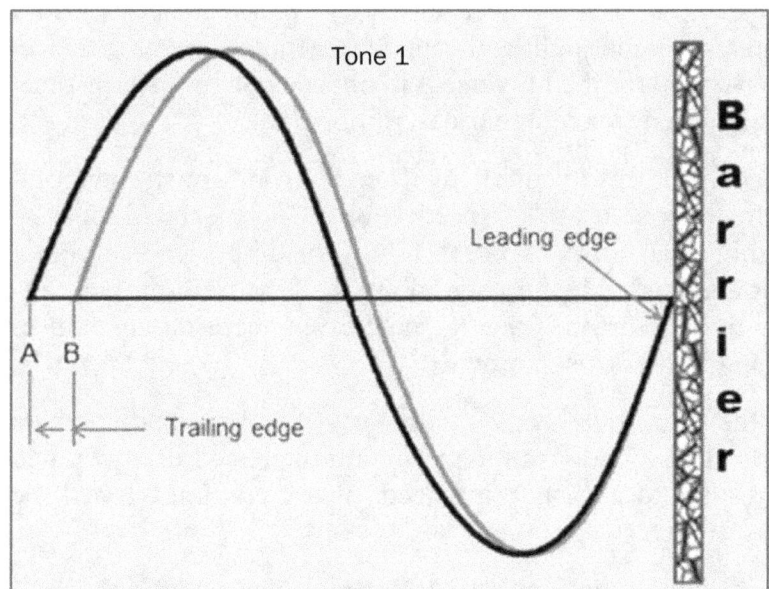

To keep things simple, we're using an old train with a single note whistle. Say a train is approaching you head-on at 90 kph. The engineer pulls the cord. The whistle blows. The leading edge of the first cycle leaves the engine heading towards you.

For the full explanation, we need to know how far the train moves in a fraction of a second and how far the first acoustic cycle has moved in a full second.

Tone 1 indicates what happens when that train is traveling towards you with a whistle frequency of 512 Hertz. The graph is a sine wave representing the air as it gets compressed and decompressed slightly as the wave begins its journey, the higher the amplitude the more dense the air. It crosses the horizontal line which represents ambient air and then goes negative to indicate less density. We're talking one cycle only.

Position A is when the engineer pulls the cord. The whistle sends the leading edge of the wave away in all directions. We only care about what's headed our way, so we slice that portion out. If the engine were not moving, the wave would appear as the image shows beginning at A, but the train is moving. In one split-second, the whistle's leading edge has moved forward, but in that same split-second the train has also moved toward B. A lot happens during that first period. The mechanism generating the sound continues its motion as the wave forms. Each portion of the wave moves closer to its leading edge because its speed has maxed out. At the end of the time period, the length has shortened by the distance the train has moved. A shorter length means a higher frequency.

Recognize that the process creating the tone is continuous. It may seem fast, but it is not done in an instant. Sound waves do not jump off the generating mechanism in the form of cycles. The device creates them, bit by bit, nonstop until some action turns the apparatus off. This concept will be more important when we consider light waves. They are also generated over distance but at a much smaller scale—billionths of meters instead of thousandths of meters.

Most references use time, but for our purpose, we need to express that distance is also a representation of time. And we must sway that light and sound are generated over distance. Most programs presented on television about the universe will always mention space-time. Since space implies distance, we will use distance in our description.

The barrier is the speed limit of sound for this material, air. The wave doesn't stop there. It continues to propagate away from its source at the speed allowed by the medium. The above depiction represents everything that happens in the first acoustic cycle. Everything shifts to the right to make way for another cycle, another period, and the train moving farther along its tracks. An important thing to grasp is that the next cycle begins at position B, and repeats for each cycle for as long as the whistle blows.

You probably have good grasp of the Doppler Effect at this point, so if a little math bothers you feel free to jump to Chapter 11. But if it doesn't, and you are interested in how much tone or frequency changes with respect to a change in speed, please continue. It will be worth it.

A little arithmetic is necessary to set this up for a standard day in air because the speed of sound varies a little with temperature, atmospheric pressure, and density. The standard day under consideration is 29.92 inches of mercury at 59 degrees F plus density and viscosity.

Speed of sound in air = 342 meters / second. An alternate spelling is metre used by our European friends.

For our discussion, we need to know the wave length of the whistle and how far the train moves in one split-second. The frequency is 512 Hertz or to some of us older guys 512 cps (cycles per second). The formula for finding the wave length is,

$$\frac{\text{speed in medium}}{\text{frequency of tone}} = \frac{342 \text{ m/s}}{512 \text{ c/s}} = 0.66796875 \text{ meters/cycle}$$

or 667.96875 millimeters. Another reason for the greater precision is because this value is used later when determining how much the train has gained on the cycle's leading edge. Further, the tags are in place for those who may ask, "Where did the seconds go after dividing?" They cancelled leaving meters per cycle where cycle is usually dropped from the conversation. Cps or c/s was used during the calculation instead of Hz to gain a bit more understanding (no disrespect to Heinrich Rudolf Hertz who was the first person to provide conclusive proof of the existence of electromagnetic waves.) The lower case Greek letter lambda λ or upper case Λ is used to cut down on typing in future references to wavelength.

Now that we know the wave length, the next task is to find how far the train moves during one cycle of the whistle tone. Ninety kilometers (90,000 m) per hour must be divided by the number of seconds in an hour (3,600).

$$\frac{90000 \text{ m}}{3600 \text{ s}} = 25 \text{ m/s}$$

The whistle (on the locomotive) is traveling along the tracks at 25 meters per second. But how long must it move at this speed before the next acoustic cycle begins? That is to say, how much time does the train have before the next cycle interrupts the process? For that answer, it is necessary to determine the time it takes for one cycle of the tone to finish. The tone is 512 Hz. One cycle takes

$$\frac{1 \text{ c}}{512 \text{ c/s}} = 0.001953125 \text{ s or } 1953.125 \text{ μsec to complete.}$$

That's 1953.125 microseconds or millionths of a second.

Therefore, the distance the train moves in that same time period is

$$25 * 0.001953125 = 0.048828125 \text{ m or } 48.828 \text{ millimeters.}$$

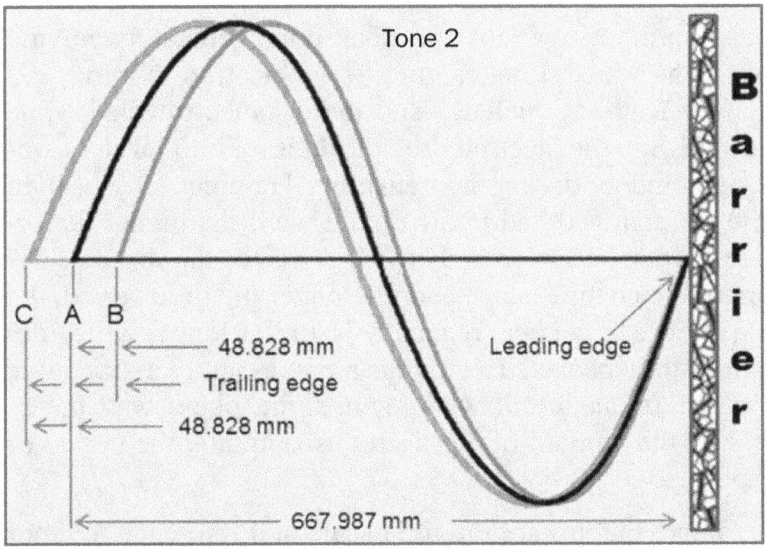

The letter A in Tone 2 points to the trailing edge of a cycle of 512 Hz while the train is at rest, and B points to the trailing edge of the same cycle at the end of one period when the train in moving at 25.000 m/s. The length of A to B represents the distance gained by the trailing edge for that one cycle.

Wave length of the new tone is 667.987 – 48.828 = 619.159 mm.

The new frequency is found by dividing the speed of the wave by its length, that is,

$$\frac{342 \text{ m/s}}{0.619159 \text{ m}} = 552.362 \text{ Hz}$$

As the train approaches, the tone is higher—approximately 552 Hz. As it buzzes by, the tone is the original 512. As the whistle moves away as marked by C in the above image, it stretches the wave by the same amount it had squeezed the tone earlier. The sound is lower at approximately 477 Hz. without showing the arithmetic.

We used the easiest parameters for our demonstration to avoid complications. The train was coming head-on, passing at 90 degrees, and then receding at 180 degrees. We were at its six-o'clock when it disappeared.

Since our subject is about how sound changes in frequency with speed, we will not go into its other attributes like harmonics, beat frequencies, or heterodynes. Please check out these interesting characteristics on your own. The internet is a great place to visit for those subjects.

There is another property of sound to consider. We mentioned barrier earlier, but that barrier is only good for that particular medium. If two people are carrying on a conversation inside a Concord cruising along at 1320 miles per hour, they still hear each other fine. So you might say that that sound is traveling at 1320 + 765 = 2085 mile per hour. You would be right. But air inside the airplane is traveling at 1320, so the sound is still only going 765 in its medium.

Wavelength of light varies exactly like that of sound except at a blinding speed. And rather than a tone, it is the color that changes. When comparing light to its acoustic cousin, a red hue would be like a bass note while a blue hue would be like a treble tone.

Light emitting electrons do not change orbitals in an instant. Like sound generators moving along tracks, light generators move along space. It takes time for an electron to move to a new position closer in or farther away from its nucleus. And the distance traveled while emitting a packet of light is measurable. Say the electron drops from level 6 to level 4 skipping level 5. It will only emit that light energy during the transition. The frequency of light given off by any electron transition is cataloged and known quite well: the higher the orbital, the greater the energy and the corresponding frequency; the greater the distance between the orbital transition, the longer the change takes, and the longer the time period, the more energy stored in the packet. That is, the higher frequency last a lot longer which delivers much more energy than a shorter time packet. Two things are working to assure more energy: longer time period, and higher frequency. It works something like a watt meter. Utility companies charge customers by the amount of electricity used and by the period of usage. It's known as the kilowatt hour.

The only things that change in the above picture are the time scale and reference to color instead of tone. The time scale is usually in nanometers or some part thereof. This subject will be addressed in greater detail later.

Please don't hate the author, but the previous text was intended to provide the WHY and HOW frequency changes with speed and location. This is a bases for understanding light waves later. There is a much easier way to calculate the change in frequency (audio only; not light) but doesn't convey much of the why's and how's.

It is
$$f = \left(\frac{v_s}{v_s - V_o}\right) f_o$$

where v_s is speed of sound, V_o is velocity of oncoming object, f_o is original frequency, and f is new frequency. Please visit https://formulas.tutorvista.com/physics/doppler-shift-formula.html for other observer relations.

Chapter 11

Freefall vs Momentum vs Centrifugal Force

Consider three types of acceleration:

1) Acceleration due to superforce and flux transparency (freefall)

2) Acceleration due to angular momentum

3) Acceleration due to externally applied force where *force / material in object = a*

Everyone has experienced number 1. Just falling down as a kid is freefall. Bumping into the ground was just the sudden stop brought on by rapid deceleration. Jumping off a couch or bed counts too, or a high porch. At such short distances, air friction doesn't count. Jumping off a three story building is something else. While air friction is not a problem, time in transition is. Since the increase in velocity is directly related to time, it means the longer an object falls, the faster it will travel. Transparency flux opposing inbound vectors bring about delta forces, and that pressure differential applied to an object over time leads to acceleration or freefall in this case—the greater the transparency, the greater the opposition to slow the acceleration. The converse is also true, the less the transparency, the less opposing forces resulting in much greater acceleration.

Earth's transparency flux opposes the inbound flux which results in an acceleration of 9.8 m/s^2 at Earth's surface. Jupiter's opposing forces create a delta force strong enough to accelerate any object at 23.6 m/s^2 while the result of Mars' and Mercury's delta forces are almost equal at 3.77 and 3.76 m/s^2 respectively.

Most everyone has experienced acceleration number two as well. Most kids tied a string to an object and twirled it round and round their heads until it extended outwardly almost parallel to the ground. Centripetal acceleration caused that. Centrifugal force (outward) is the result of acceleration toward the center. . . . Well, maybe not the newer kids.

Everyone has experienced number three also. Any moving vehicle will apply acceleration to a body attached to it. It will attempt to force the body through its seat rounding the bottom of a hill or toss the body off the seat when topping the hill. When rounding a curve quickly, both two and three are witnessed at the same time. If the body gets thrown off the vehicle over a cliff, we're back to number one. That's when you want to be on Mars.

Freefall: No force is felt during freefall because every atom of every molecule in a person's body has the same net force applied. A person feels something only when stress or strain to some part of her body is applied. An accelerometer can't detect freefall because every atom of every molecule of the device has the same force applied. See Accel 1.

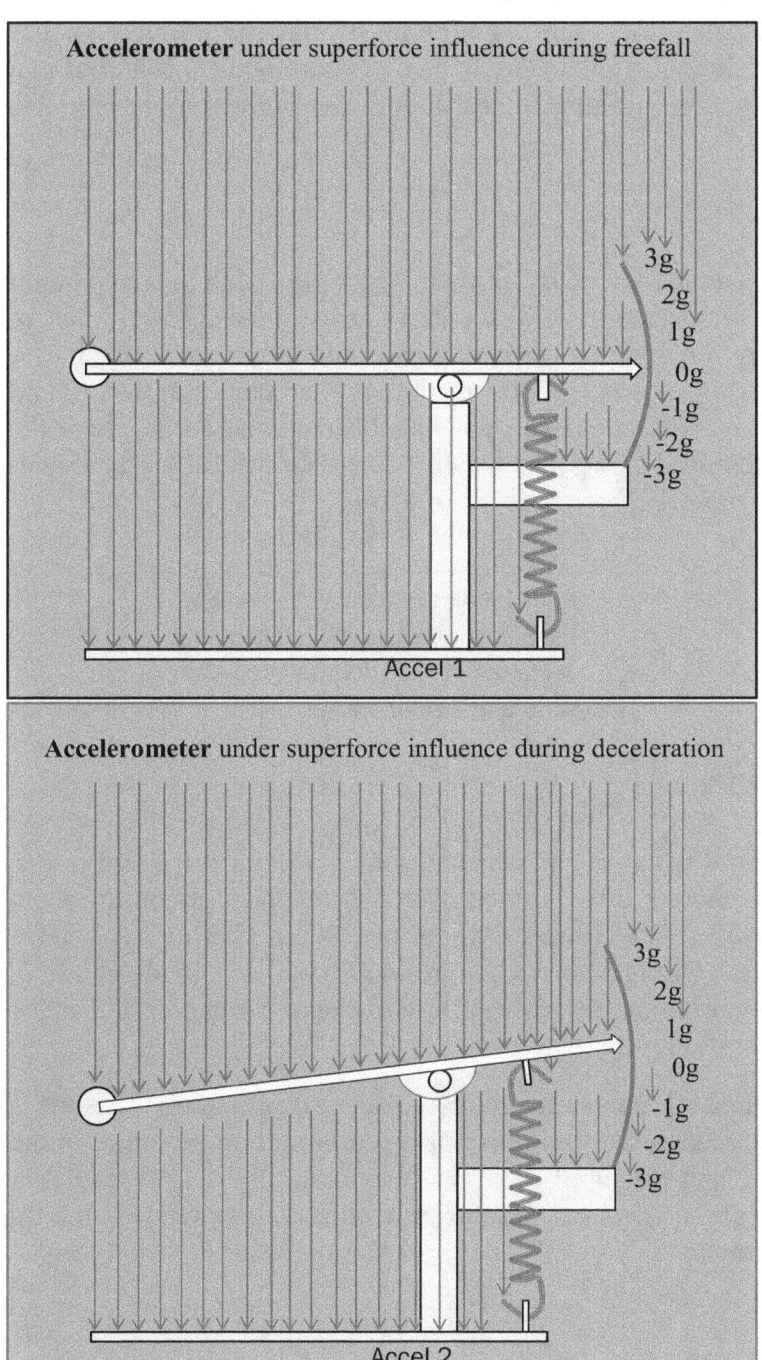

The image contains only earth-bound flux for clarity. It does not show counter force flux exiting the earth due to transparency. Every molecule is subjected to the same net force driving the entire assembly toward Earth's center. Since the delta force is continuous at the assembly's current location, it must increase in velocity over time, which results in acceleration. The force changes as the assembly moves closer to Earth's center where they all become equal. The universe loves equality.

Accel 2 depicts an accelerometer sitting on a desk, sitting on Earth, which can no longer move toward Earth's center. Earth as a solid object, resists any further movement of another solid object by applying an equal force against the assembly's base. Other ridged components convey the resistance through load transfer mechanisms. The counterweight and lever are free to rotate about the pivot point, and they contact the earth by way of that single point. The net force acts on those parts as well but with restricted movement by the spring.

Acceleration forces due to changes in travel direction respond to a different set of actions.

Flux transparency of the targeted body determines the amount of acceleration a falling object undergoes. Once inside the acceleration zone, the amount of material a much smaller body contains doesn't matter when it comes to closure of two or more bodies. Material in a falling body near Earth's crust is ignored because the earth contains billions more particles than the smaller object.

Acceleration Zone

Anytime a small object is greatly overshadowed by a much larger one, such as a person on the edge of a sky scraper compared to the surface of the earth bearing the sidewalk below, the smaller object is in the acceleration zone of the larger one. The larger shadow encases the smaller object completely. All the vectors opposing the smaller object are outbound from the larger. Inbound vectors overwhelm those opposing because the earth filters the outbound flux. This affects a differential force that would accelerate the little guy downwardly should he lean too far over the edge.

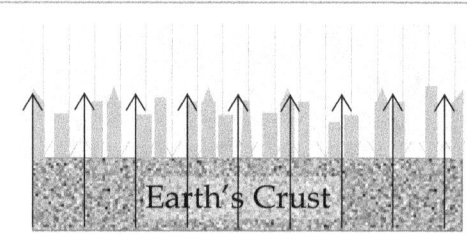

Remember, the darker the shadow, the less the resistance to oncoming vectors. Here the darkness is represented by bolder arrows.

Notice here that we have emphasized the cause of acceleration as a delta force and not as gravity. When the vectors leaving Earth's crust are subtracted from those incoming from our Universe's edge, the net force is a -9.8 Newtons where the minus sign means headed towards the sidewalk. If the ledge is 1600 feet above the ground he will accelerate at a little over 32.2 ft/sec for 10 seconds. Ignoring air resistance, he will make contact at 322 ft/sec or nearly 220 miles/hour. . . . Be careful up there.

Now, that -9.8 Newtons is not pressure even though it will create pressure. Pressure is force/area. We have used them interchangeably for ease of discussion, but they are not the same. Say the force is 100 pounds, and it is applied over 10 square inches. That results in 10 lbs/in^2. We'll have more to say about this later, but for now, that -9.8 Newtons is the total net force on every molecule of matter.

Chapter 12

Orbits

It is said that anything in orbit is in freefall. The alternative view will modify that statement.

As mentioned before, a body in freefall is accelerating even though there is no apparent marker to identify the acceleration. Further, some scientists declare that bending of space itself is what regulates orbits. That is, the body is in motion and simply follows the curvature of space surrounding another object. Again, we can see the answer with a few graphics.

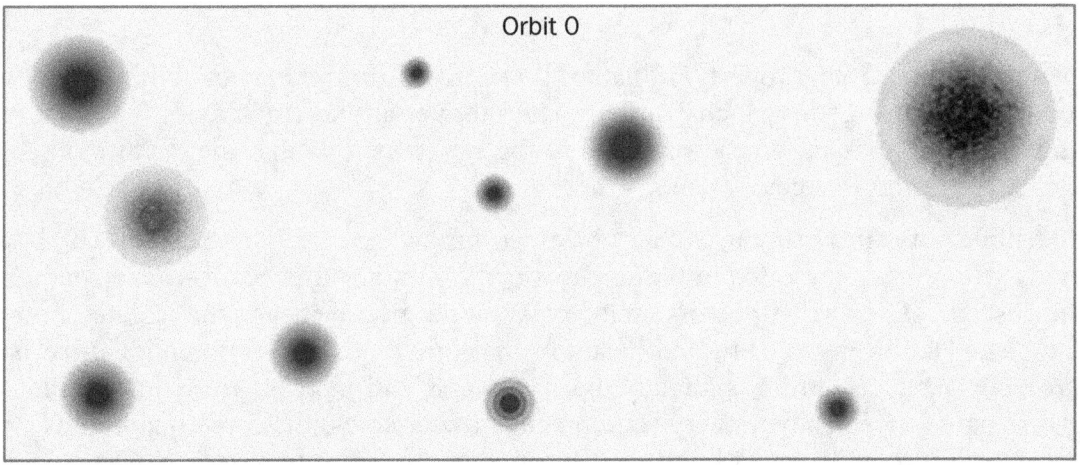

Let's look on the negative side a moment. Since an object's transparency shadow is a beacon, and all other bodies have such and respond to that beam, the power of those beacons determines all movement within range: the greater the power, the greater the range. Beacon is in the negated sense. It's almost like an astronomer's negative where the stars are dark and the surrounding space is clear. Except they vary in darkness from almost nothing to very black. Each one is also much lighter at its edges than at its center because of the funnel effect. Actually, flux that makes its way through a heavenly body with an attached light beam stands in opposition to any object it hits, and those vectors that do not make it through, do not. That probably can go without saying.

Each object's signal is visible to all other bodies in every direction. The arc size is dependent only on their distance apart and physical size.

Every object in Orbit 0 is forced to all other bodies. While the motion may be complex, it still exists. Each body is accelerating toward some point centered by the summation of all objects' transparency shadows. Although accelerating, they are in freefall. And at the same time, momentum is driving each object forward, onward in whatever direction the original impetus intended. Thus comes a battle of two dominant forces of our Universe: one under influence of the remaining superforce; the other, a law laid down in the very beginning.

This is an oddity: every child's dream is their discovery of perpetual motion where something can continue moving with no input. We soon learn that that discovery will never occur. However, as far as the universe goes, everything is in motion. Nothing is at rest. To those dreamers of long ago, perpetual motion exists. Perpetual stillness does not.

All matter was born in motion because the composite energy from which it came was expanding. That was the initial impetus, the original motivation, and by law, it will remain so. Any change in direction and velocity of that matter shall come about only by external forces.

Momentum is an internal property of matter. It is derived from the amount of material within a body, and a change in velocity of that material over all increments of time. That is, the amount of matter making up the object, and the movement of that matter through spacetime since the birth of its components.

While trillions of miles from any other object, so far away there is no beacon within sight, any body will remain in motion at the same velocity until a single beam from a shadow hits it. The absence of superflux vectors creates that single ray. It means the isolated body has detected a reduction in pressure and heads in that direction. The differential force is tiny but present. The acceleration is already underway and will increase over time as the shadow grows darker. Remember, every time the shadow doubles, the force quadruples. When referring to closure, millions of lightyears become thousands of lightyears rather quickly.

Matter has built-in goggles that see incoming flux as Earthlings see sunlight. When something blocks that light, earth folks see a shadow. It may be so small their eyes cannot detect the occultation, but it is present. Objects of matter are more sensitive to flux shadows than humans are to light shadows; therefore, they react by moving toward those shadows. Say a small negated beam from far away strikes an object. That single beam means that that star is blocking out one or more background vectors in turn causing a difference in pressure.

The ongoing struggle between momentum and the superforce causes an object to react to the slightest force differential by moving it in the direction of least resistance. The result is a change in velocity, and that change is known as acceleration: the greater the change, the greater the acceleration.

Generally speaking, a body's gravity is presented as its acceleration be it Earth's, Mars' or Jupiter's. But acceleration is an effect, not a cause. The cause is a delta force. Many authors declare that there is no such thing as the force of gravity. Others declare that without acceleration, there could be no force.

This alternative view believes that the only remaining superforce from composite energy is responsible for pressurization. Transparency is responsible for acceleration. Both must exist at the same time to bring about movement. That is, an unequal force applied on one hemisphere causes movement. Over any given time span, a net force continues to increase an objects speed. The span may in microns per $\sec^2$ or megameters per $\sec^2$. That is the default. Any differential force is always applied over a given time period. It may be short or it may be long, picoseconds or teraseconds. The instant the force disengages, the body will continue on its path with all the previous energy stored as kinetic energy. Admittedly, sometimes the body may come to rest in contact with another before the force uncouples.

The period to make a complete stop also covers a time span, and sometimes it is in the nanoseconds range. That's when all hell breaks loose; the object releases all the energy in a very short time. It's something like a small pressure over a great area becoming a large pressure over a small area. It's the same thing.

During acceleration, a body stores the force responsible for its change in velocity within itself as work energy. The force is doing work while accelerating the object. The longer the force is applied, the greater the acceleration, the greater the acceleration the more energy stored in the body. Kinetic energy is the summation of all changes in velocities—the summation of all acceleration that each piece of matter within has made since its creation.

After adding up all the changes in momentum, that is changes in velocities, ($\rho=mv$) we get

$K = \frac{1}{2} m v^2$. Where K is kinetic energy.

Even if an object is moving slow, at some point in its history it had to have had a change in velocity to arrive at its current rate. And each time a force has been applied during that history, the entity underwent acceleration.

Simplistically speaking, $v_0 + v_1 + v_2 - v_3 \ldots$ and on and on, sometimes positive, sometimes negative. The unit of mass records all changes in v, stores them, and at the proper time, releases some or all the energy as required by law.

A student of calculus can glance at the formula and realize that momentum is the derivative of K with respect to v. It also means when $\rho=mv$ is integrated over time with respect to v the result is

$$K = \frac{1}{2} m v^2$$

thus semi-proving the previous statement.

Momentum is also a vectored quantity. It has direction because velocity has direction.

Until this chapter, all examples have implied objects coming together on a direct path. However, while that does happen, it only makes for one object becoming larger, or sometimes making smaller objects out of larger ones. Another approach is objects on an indirect route that result in capture and one going around the other. The one with the most material becomes the host.

How the superforce changes an object's forward motion to a circular one is complex, but here goes. Sorry, but the explanation requires more images.

While trying to avoid the term gravity, sometimes it is easier to give in and use it when applicable. In this case, it is required to find the center of gravity of two objects on a seesaw. Below is that representation. Torque is the result of a force applied to an object at a given distance from its fulcrum if the terminal is fixed. If not, the result is acceleration.

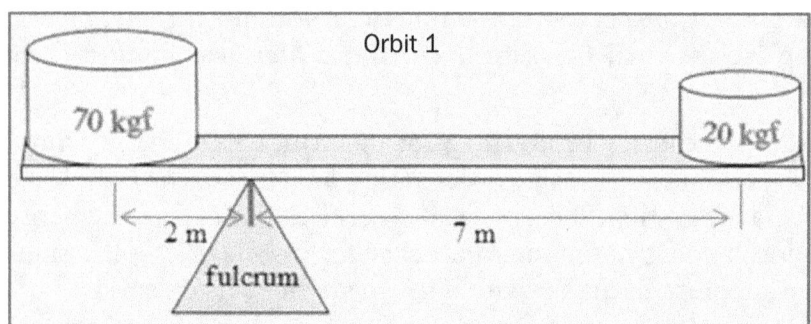

Our Universe's laws have been setup such that they follow mathematical principles, especially the product of properties of its objects. Both containers of Orbit 1 are filled with concrete. However, they may each rule over their domain equally. To maintain the system's balance, the weight of the smaller multiplied by its distance from a point must equal the weight of the larger multiplied by its distance from the same point. That is,

$$7m * 20kgf = 2m * 70kgf$$

where kgf is kilograms force as opposed to kg mass. The result is 140 m-kgf torque. Any variation of that number for either side will induce rotation around the pivot point. Americans are more familiar with ft-lbs of torque, and later on we use the term $P1*V1$ where there is a pressure related to volume.

Should someone lift or push down on either end, the added force will induce an unbalance and rotation will commence. When the pressure is removed, the system will return to a level position.

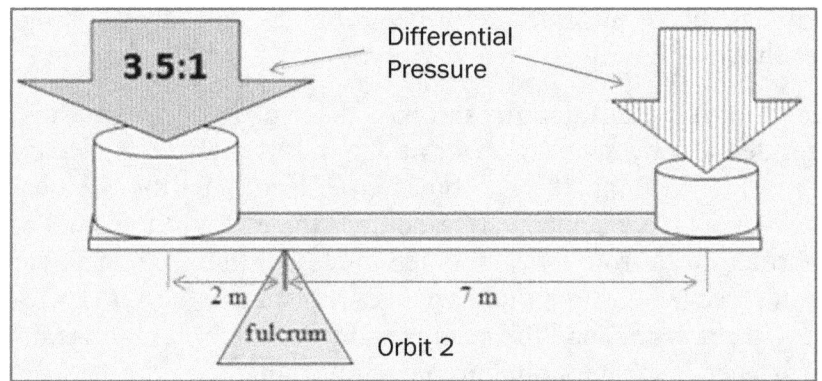

However, that procedure requires an inference of mass to obtain the weight of each object of which the *Alternate View* is attempting to remain silent. The image in Orbit 2 replaces mass with net forces that act in the same manner.

The image of Orbit 2 results in the same weight but without inference of mass. The weight is equal to the differential pressure brought about by each's transparency index. There are 3 ½ times more matter in the larger object meaning flux interacts with that many more atoms of various makeup. Verify that by multiplying 3.5 times 2m resulting in 7m which equals 1 times 7m resulting in the same value of 7m. A mystery appears though. We are multiplying 7m by 1 ? in one case and 2 m by 3.5 ? in the second case. Since there is no current identifier describing differential pressure, we can simply use pounds or Newtons as force. After all, the result of the superforce's action creating a differential force is what determines the balance or imbalance of the system under consideration. Currently mass*acceleration=force which is pounds in America and Newtons in most other countries. For this alternate view, it is just force without any multiplying that results in the same weight, pounds or Newtons.

The same goes for objects in orbit. Distance is the equalizer, and impetus is that extra force applied to induce motion around a point. A small planet can create movement of a giant star. Only a teeny bit maybe, but still it can cause a reaction.

The struggle for equality of objects on a seesaw is small versus objects in orbit under the influence of the superforce. Yet that other powerful force is presented when discussing orbits, momentum.

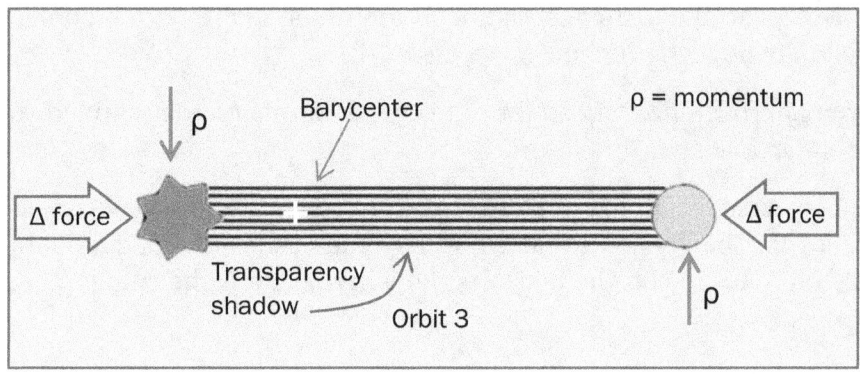

Momentum is a variable property of matter. As said before, the net force instills it within an object; the longer that force is applied, the greater the velocity, and the greater the velocity, the greater the momentum. Once that force is removed, it ceases accelerating and continues with its last bit of information. It keeps on doing what it's doing. Its energy of motion guarantees that it will.

The center of force of a teeter-totter is at its fulcrum, and the center of force of an astronomical system is its barycenter denoted by a + sign in Orbit 3 and 4. A fulcrum is usually a fixed point and a barycenter is free to float. If the orbit is an ellipse, then the distance from center varies greatly. However, as the distance varies, so does the effect of the smaller body. That is, when a body is farther away from its barycenter, the less influence it has over its companion. With a change in force proportional to a change in distance, the system remains intact. In this case, replace force with ρ and the orbit is stable. As an object slows, its momentum is reduced but the distance is greater maintaining equilibrium.

Perhaps a seesaw version would be a pail of water that leaks. As water drains, the distance from the apex must increase to keep things equal, or the distance from the apex on the other side must decrease to keep the numbers equal. At some point the leak must be repaired or the system accepts the empty bucket as a fixed unit; if not the system rejects the bucket and becomes unbalanced.

That's what happens to a comet. At some great distance from the sun, the leak stops, and the comet slows because the superforce overcomes its momentum. The weight becomes stable, the cycle begins again, and then the leak resumes as ice melts when the comet comes closer to the sun. Sooner or later, all the ice melts and the only thing left is a large rock with a relatively stable orbit. Or not.

On Earth, a falling body will accelerate at 9.8 meters per second per second. This we already know, but how about a comet out near Neptune? It will accelerate toward the sun at a varying rate as described earlier when discussing powers of two. That's because as the comet gets closer to the sun, the delta force increases as the object's radius grows. Two orbital factors make a comet's path complicated to describe: (1) the acceleration varies as the comet comes closer to the sun because the closer it comes the stronger the influence of the sun's transparency shadow, and (2) the comet loses material as it approaches the sun, thus creating a lighter shadow. However, the sun contains much more material than the comet, so the loss of comet material may be insignificant.

Our simulation will not even approach how those far out objects' orbits are determined. It will be a simple planet circling a lone star.

For most readers the versin, or versed sine, may be unknown; however, it dates back several hundred years. Table values for the versin are rare, but they can easily be found by subtracting the cosine of theta from one: $1 - \cosine \theta = \versin \theta$ where theta is the angle under question.

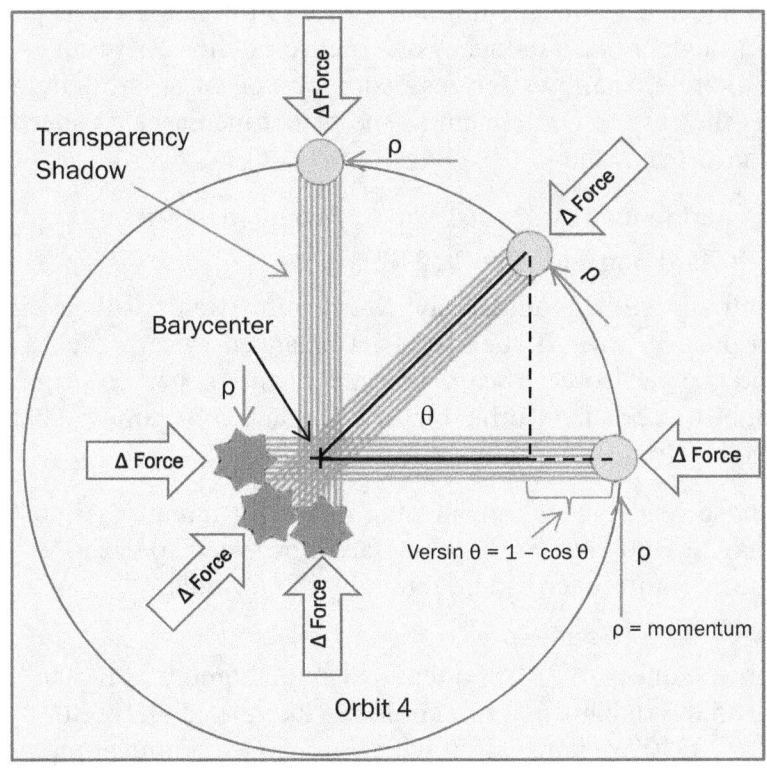

We bring up the versin because it relates to a change in velocity brought on by the object's transparency shadow. Our model will be simple but yet reveal visually how an object accelerates in orbit the same as if it were on a straight collision course with its host.

Orbit 4 shows a small astronomical system overlain on a compass of a unit circle. Beginning on the right side of the circle, we see a small planet with its transparency shadow extending past the barycenter to a red dwarf. The red dwarf's shadow overlays the planet's, and together those shadows emulate the beam of a seesaw. While invisible, the shadows mechanically connect the two objects.

The numbers chosen for this example are not arbitrary. They were calculated such that a whole number could be used for the orbital speed of the planet. That speed is 72 meters per second.

radius	circumference	orbital period	speed
990071.07 m	6220800 m	1 day = 86400 s	72 m/s

Universe's Mathematics

The values in the tables have been squeezed into the unit circle of Orbit 4. We used trigonometry values based on the unit circle that has a radius of one metric of choice. By superimposing our orbital system on the circle, it allows for easy computation of an arc length and an x axis value. The x value is the eastern component as the planet increases its speed to the left; the y component is the northern portion.

At position zero, the velocity (V) is northbound at 72 m/s. The x, y components are

$$x = 0, y = 72 \text{ giving } V(x, y) = V(0, 72).$$

The battle between momentum and superflux is underway. Momentum wants to keep it going straight, superflux wants to push it towards the star. In a balanced system such as these two objects, momentum and the net force toward the sun result in just the right amount of acceleration of the planet to keep it in orbit for many, many long-times. That phrase is a tribute to a friend, Su Mann Ho.

A short explanation follows for those readers who are not familiar with radians. Almost everyone is familiar with a circle having 360 degrees, but there are other ways to describe a circle's properties. For example, the circumference can be found by the formula,

$$c = 2*PI*r \text{ or simply, } 2PIr$$

If the radius is one, the circumference reduces to 2PI, and that's also the number of radians in a circle. That is, there are 6.283185307 radians in a full circle. So a circle can have 360 degrees at the same time it has 6.283185307 radians, all of which are angles. For example, a quarter of a circle is 2PI/4 or PI/2 radians or 90 degrees. The length of any portion of its circumference is also known as the arc length. Arc length is usually identified as

$$s = r*\text{angle in radians, normally seen as } s = r\theta$$

where r is radius, usually in meters or kilometers, or some other appropriate term. Pictographically speaking,

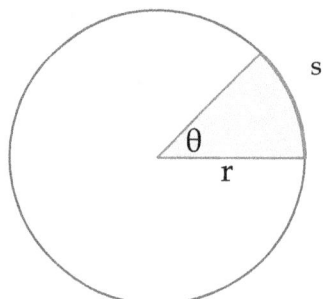

Say the angle above is PI/4 = 0.785398163 and r is 2 meters. The length of arc s is 1.570796327. It may seem like an incredible amount of precision, but sometimes it's necessary to achieve the proper goal, especially when dealing with millions of kilometers.

We already know that acceleration is a change in velocity over a given time span. This example is going to show that change over a large time and distance beginning at angle zero and ending at angle PI/2 after passing through PI/4. In that distance our object's eastern component of its speed will vary from zero when it's going north to 51.91 m/s at PI/4 to 72 m/s when it's headed east at PI/2. The northern component is reduced to 51.91 m/s at PI/4 from 72. At that position, the direction of V is north-east @ 72 /s.

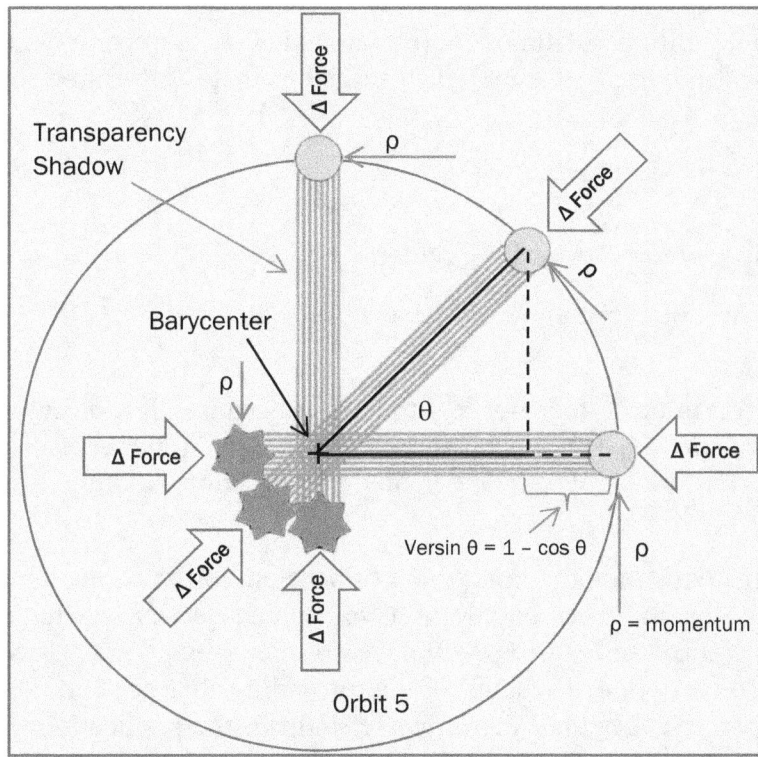

Orbit 5

We reproduced the system as Orbit 5 to keep from flipping back and forth.

At any angle, the net force is always compelling the planet to change velocities which results in acceleration toward the barycenter. While some authors insist it is a change in direction that invokes the acceleration, this alternate view is that the change in direction is an effect not a cause. The cause is acceleration. So, acceleration can be both, a cause and an effect.

With each displacement, the delta force is always perpendicular to the momentum forcing the planet to turn toward the barycenter.

In reality, this example does not show the acceleration toward the barycenter. It only shows that the change in velocity goes from zero m/s east to 72 m/s east over a long time span of 21,600 seconds. The same thing happens if we begin at PI/2 and continue to PI. We could do this all year, but the only thing we would understand is that the velocity slowly changes from zero to full stride every 21,600 seconds. However, from this demonstration it is easy to understand that the superforce brings on a change in velocity which means it is also responsible for accelerating the planet. But the acceleration needs to be directed toward the barycenter instead of just to the planets left side.

For that, we need to chop the circle into ever smaller chunks, slices so small that the acceleration approaches the direction of a line perpendicular to the tangent. A line normal to the tangent goes through the planet and the barycenter. Furthermore, the superforce must also force a change in momentum. Momentum has direction, and since traditionally ρ = mv that means a change in velocity accomplishes a change in momentum.

If the simulation is performed one million times per revolution, that is every 1/1,000,000[th] of a revolution, one can visualize how the direction of acceleration moves closer and closer toward the barycenter with each added cut. Hopefully this demonstration gives insight as to how acceleration takes place during an orbit with a perfect circle.

When the host of an orbit contains millions of times more matter than its companions, a companion's speed defines the orbit ignoring its material. It turns out that the formula for centripetal acceleration is

$$A_c = \frac{v^2}{r}$$

with no mention of matter. A_c is acceleration toward the center.

For a great explanation of how the formula is derived, see the PhysicsLAB at http://dev.physicslab.org

As explained in chapter 11, a person in orbit does not feel the effect because all molecules in her body have the same pressure applied. There is no stress even though all her enclosed atoms are undergoing acceleration. It does not matter whether the direction is straight or curved.

We are not challenging Doctor Einstein's thought experiment of acceleration in an elevator in deep space, but his imagined motivating force could not have been caused by Newton's idea of gravity. Acceleration due to flux vectors or gravity have no internal markers. There is no stress on any part of that body, period. Einstein's force providing that acceleration would have to be developed by power other than the universe. Indeed, the application of flux vectors will provide the same impression of curved space as Einstein had envisioned.

We earthlings feel stress when standing on the ground because all molecules are trying to accelerate, but earth's crusts says through the feet, "This is a far as you go!"

So with all that said, we only feel the strain of the applied force while not really moving along with it.

Chapter 13

Speed of light

Most everyone knows the speed of light is just under 300 million meters per second or 186,282 miles per second. But, how does light travel at all? Why is it a constant no matter how fast the emitter is traveling or in which direction? Most folks shake their heads at the very thought of such a nonsensical thing, and rightly so. Let's see if we can make sense out of it.

Let's learn how light travels in the first place, and for that we must revert back to mass which is an effect of the superforce.

Strike the idea of comparing light with matter. Light is energy; it has no mass. It only has an equivalence of mass. That is, so much matter is *equal* to so much energy.

A ball is a good example when discussing matter. If a baseball pitcher is riding along in a boxcar traveling at 70 mph and throws a fastball at 96 mph, the baseball is traveling along the track at 166 miles per hour. But, it hits the end of the boxcar at 96 miles per hour for obvious reasons: though not with electromagnetic energy (EME). As we argued before, the superforce is a flux, and it has an affinity for electromagnetic waves.

Image of electron changing state

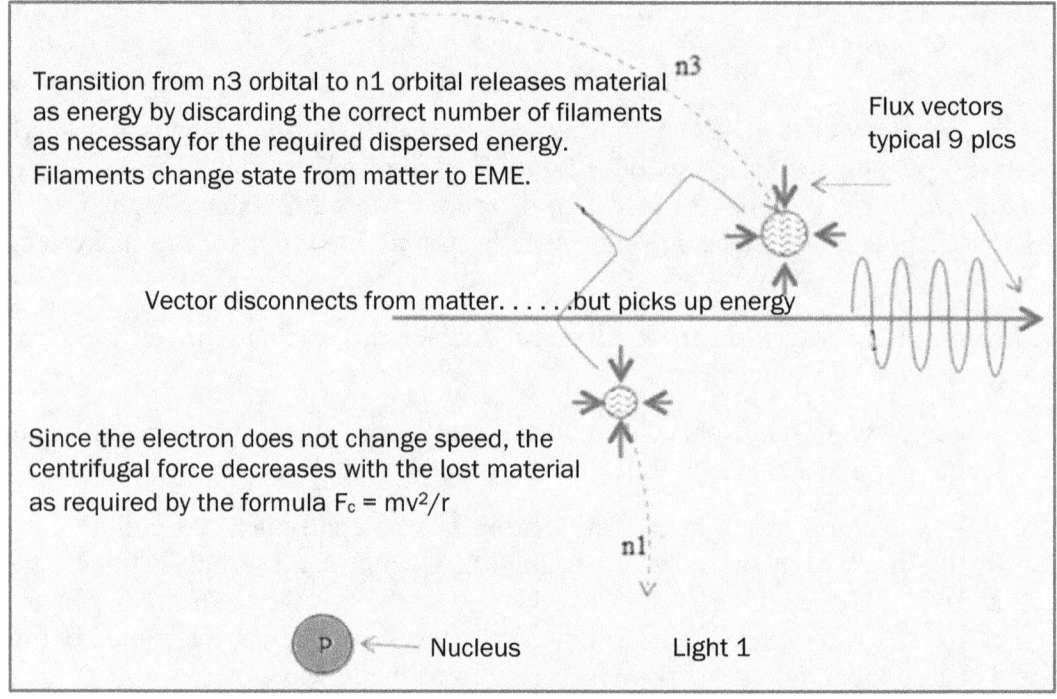

Remember, superflux applies equal pressure over an entire particle. Let's say it's an electron. When an electron transitions from a higher orbital to a lower orbital, it must lose a small quantity of matter. Light 1 shows how some vectors slip off the particle as it becomes smaller during the transition. And as it becomes smaller because some of its matter changes to another state—energy, the vectors may seem to point to emptiness. But take a second look and notice the packet of energy now in place of the disappearing matter. Each vector is now attached to electromechanical energy—light. Superflux loves to push around EME as much as magnetic flux loves to push around electrons.

We must take a side step for a moment. Superflux travels at a constant velocity, so when a particle moves from under or away from it, the vector immediately gets up to that velocity. It will continue moving at that rate until it encounters another object. We say velocity because it has both magnitude and direction, but we will go along with the rest of society and call it speed. So it seems that superflux moves along at c. In reality, light moves through space attached to the superforce. Again, light is motivated through space by superflux. Since vectors travel in every direction, light heads off in every direction from where matter changes state. For now we will identify the light as a sine wave. Others may identify it as a photon. That may make it a little easier to visualize, but we prefer energy over a particle because energy is spread over a much larger volume. It's more difficult to imagine a small particle like a photon contacting another small particle, an electron. The odds are against that occurring as often as it must. However, the odds of a large volume of energy coming into contact with a small object are fairly good since the volume of energy to matter ratio is millions or perhaps billions to one. Therefore, let's keep light in the form of a wave of electromagnetic energy.

The vector carrying the packet of energy will do so until it passes within reach of another electron. If the electron can absorb the energy, it will do so by sucking it in and changing it back to matter within its body. Let's say the sinewave is a rope. Any part of that rope can get within reach of a small electron and the electron grabs hold and gobbles it all up as if it's one continuous strand no matter how long.

Of course, the electron will gain weight, increase in size, and head out to a higher orbit. The little vector now has another object to act upon. That is the ideal situation for this vector. There are other situations. The light may collide with another form of matter. If the new form can use it, it will; if not, it won't. Nevertheless, when the vector contacts an object it applies pressure because it doesn't give way to matter. The energy packet is picked up by another free vector and sent on its way, any direction will do, until it finds another home. It

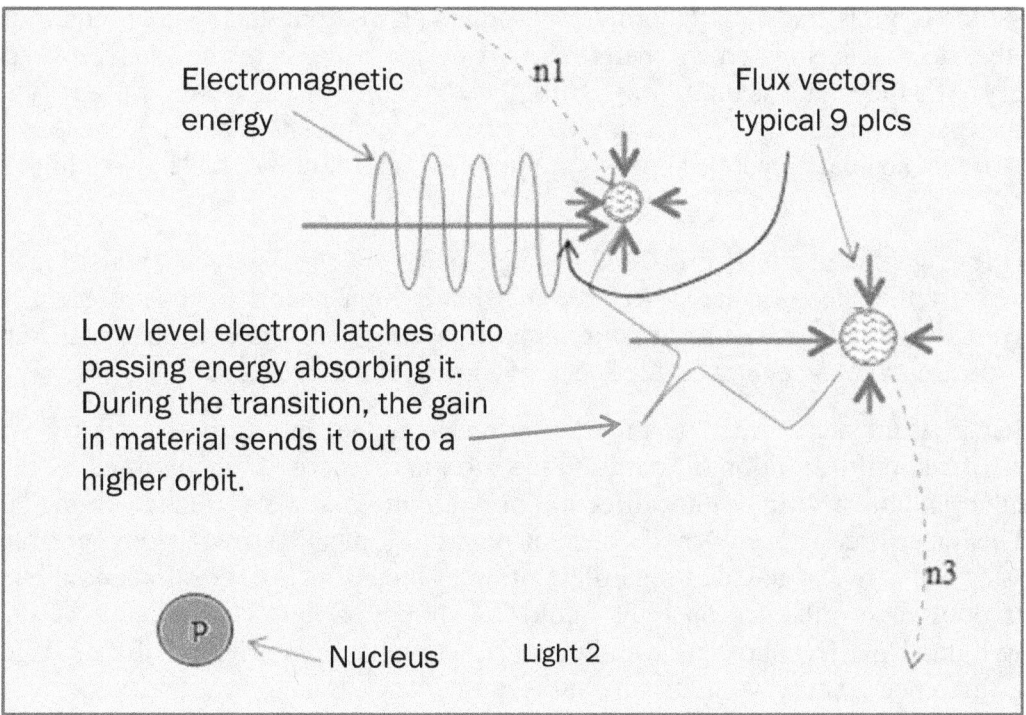

could be an eye, a telescope, or it may fall on a leaf and assist in manufacturing a little sugar. Light 2 shows the reverse process.

The new material will affect changes. If the electron is in a low state, the additional material will send it up to a higher orbital. Electrons revolve around protons at almost light speed. The only way an object can enter and exit that orbital without the electron colliding with it is for the object to be traveling near c. For all practical purposes, all orbital layers are solid for objects and impenetrable to most everything but superflux and affected only by chemical or electronic action.

Actually, the above is an oversimplification. In the real world, an electron or any other object moves through millions of vectors temporarily blocking them and then slipping by to allow the vector to go on its way. The above transition requires millions of flux vectors to complete instead of just one. That is, light is attached to all those involved. It is just easier to discuss a single vector even though millions are involved.

Say an electron revolves around its nucleus at the same speed no matter how far away the orbital is located. It can do so by gaining or losing the correct amount of material. If it gains a little material, it will jump up to a higher energy level because the speed is the same and centrifugal force acting on the larger body will send it outbound. When it takes on more energy, it gains material. Bear in mind the orbits of electrons are not like planetary orbits. Some orbitals trace out spheroids and others swirl around as if in clouds. The point is that the electron has absorbed energy carried by the flux vector and has proceeded accordingly. And the vector is on its way at the speed of superflux in search for another object or another packet of energy. Even a single electron may simulate a cloud because the speed is so great that apsidal precession moves the perihelion around the nucleus at a high rate by itself.

So, electromagnetic energy is propelled through space via superflux at a consistent velocity. It is really the speed of the superforce that binds light beams to that constant, but we won't resist the old way because it's ingrained in society too far. With that said, superflux propels electromagnetic energy at the speed of light, also known as c.

Now that is A BIG DEAL. Imagine this. Mankind no longer has to be concerned with the problem of adding the motion of matter to the motion of energy being emitted by that matter. Neither an object's motion nor direction nor velocity has any influence on the speed of light. The superforce is a universal constant regardless of one's own reference frame, regardless of one's own speed, and regardless of one's own direction. Motion and direction of emitters only have influence on light's color as shown earlier. The velocity of an object emitting light also affects both time and distance within its own reference frame. When discussing reference frames we mean the subject's context. Is it the local context or the observer's context? Whose world is being observed?

Scientifically, the only thing that changes with the acceptance of superflux propelling light around our Universe is the concept. That is, an object's speed has nothing to do with the speed of light. Its speed affects only time and distance and color and amount of material within. As far as it affecting the material within, that is doubtful, and we will discuss it later.

Since this alternate view is more about how or why something happens instead of what happens, the following clarification should make it more clear: superflux travels at a consistent speed, but its force applied to an object drops off exponentially as that object's speed increases. Removal of trailing force could actually mimic universal drag.

To make a point or more than one point, a change in energy and mass is ignored in the foregoing argument, and the term mass is used in its accepted meaning.

Chapter 14

Energy Requirement vs Light Speed

We need a special device to demonstrate how an increase in an object's speed requires much more force to accomplish the same additional acceleration. It just so happens that one exists. We just invented a frictionless constant force device, something similar to a steel tape measure, that applies a continuous force to an object. The extension/retraction maximum speed is governed at 1000 meters per second. Its contact to an object is also frictionless. A scheme has been setup that requires two such units on the next page. For our experiment, 10,000 meters should be sufficient in distance with 5 Newtons applied to a ball with 1 kilogram mass. Again, we must bend to conventional terms such as mass instead of flux transparency to convey an idea.

Since the ball has a sustained force of five Newtons on opposing sides it will remain still. However, if someone applies an additional force of 0.02 Newtons to the left side for two seconds, the ball must accelerate during that two seconds, and then it must maintain a steady speed until an opposing force stops it, or until it bangs against the end of the line. Just as in our Universe, momentum keeps the ball moving along as it should. If so desired, the reader can do the math to determine the acceleration required to get the object up to 0.04 m/s (4 cm/s) as an exercise. If it is a different answer, let us know at,

AnAlternateView.space.

In fact hopefully we'll be open for discussion of any subject in this book.

The contraption is shown in LS 1.

This may be a good time to imagine each vector as having 2D (two dimensional) attributes as space-time. Imagine when a vector is stretched, time period also stretches—time speeds. When it is compressed, time period also compresses—time slows. More on this in chapter 17, time dilation. Otherwise, vectors represent a rectangle where base is period and height is length between ticks and tocks. High is slow running, short is fast running.

At t_0, the east-west forces are equal, so no movement occurs. At t_1, a net force of 0.02N is

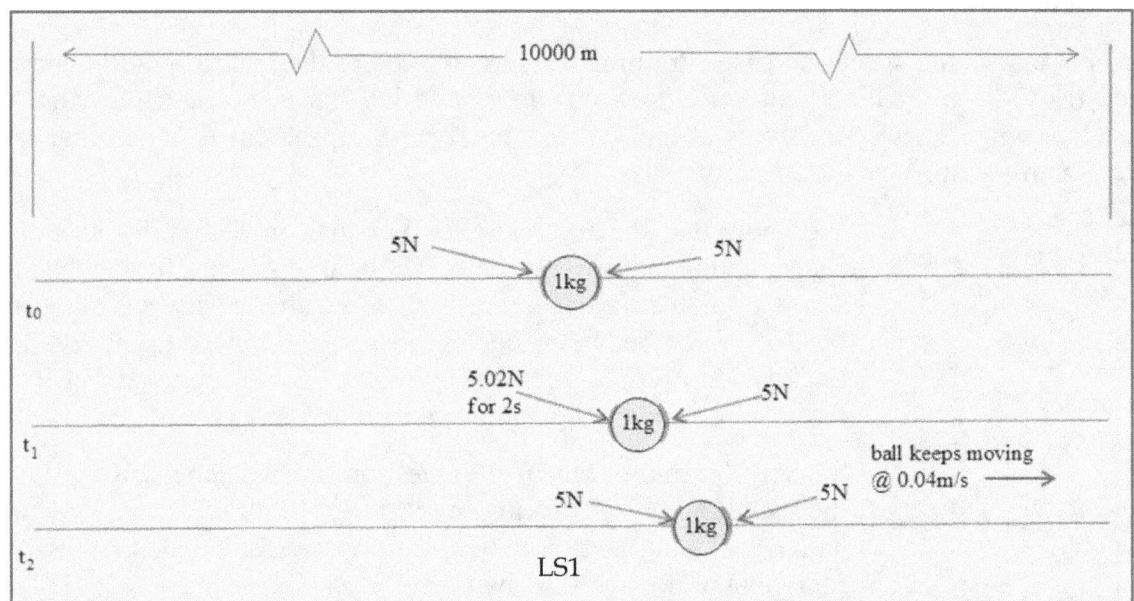

applied for a short time; at t_2, the difference is removed. The ball continues to move at 40mm per sec. It will continue that speed until something introduces an opposition force to stop it, or it will continue until it crashes into the eastern terminal.

Using the same miracle device, we can explain an important property of light: the work required to accelerate any object increases as its speed increases. It is negligibly when slow but considerably as its speed approaches that of light.

As noted above, the maximum extension/retraction speed of this unit is 1000 m/s. For slow speeds, the simulation of momentum works fine as our drawing indicates, but just as the universe has problems near light speed, so does this device. If scale is important, the proportion of 0.04 m/s to 1000 m/s is rather large at 0.004% of max speed. When compared to c, it is close to escape velocity from Earth, but it's okay for our purpose. At that rate, the left and right forces on the ball are still equal.

But what happens when something accelerates the ball enough to result in a movement of 1000 m/s to the right? At that speed, the pushing force has separated from the object, and the ball has no force on the left side. It has outrun everything that assists in movement along the x axis to the right. All the pressure is on the opposing side which results in an obstruction. The only way to keep the ball moving at 1000 m/s is to add enough external pressure to keep it moving at the new speed. Unless one can run alongside the ball at 3600 kph applying the force needed, the task is overwhelming. Let's just admit that the ball can-

Universe's Mathematics

not move at a speed equal to or greater than the maximum extending speed of the force in either direction.

It seems that as the speed increases, trailing pressure applied to the ball drops off. The device fails to keep the continuous five Newtons applied to it. That is a design flaw that we must live with. There is no workaround, so let's analyze the problem to determine how much the force drops as the speed increases.

percent max speed	applied force in Newtons
00.00	5.0000000
00.01	4.9999975
10.00	4.9749372
20.00	4.8989795
30.00	4.7696960
40.00	4.5825757
50.00	4.3301270
60.00	4.0000000
70.00	3.5707142
80.00	3.0000000
90.00	2.1794495
99.90	0.2235509

Table VT 1

We know that the force applied is equal on both sides at a slow speed, and we know that there is only an opposing force when the speed is at maximum extension. Somewhere between slow and fast, the force begins to decline. It could be instantly, it could be linearly, or it could even follow some sort of curve. In order to find out, we need to do some analysis.

Since 1000 m/s is out of the question, what happens if we boost the speed to 60% of maximum (600 m/s)? Intuitively we know the ball is not out-pacing all of the force at that rate, but it is out-running a portion of it. How much?

A virtual test shows that at a speed of 600 m/s, the remaining force exerted on the object is 4N on the left hand side. Under these conditions, Newton's first law of motion fails. The object cannot keep going at the same speed once the accelerating force is removed. It must slow until both sides have an equal force applied. If that is true for 600 m/s, it is beneficial to know how much the force drops at other speeds relative to the maximum.

See Table VT 1 for how further virtual tests reveal that from 00.00% to 00.01% of maximum speed there is no determined drop in the 5N. That test uses six decimals; at seven there is a noticeable change. However, from 0.011% to 100% the drop-off is gradual at first then increases rapidly after 60% at any normal precision. Not shown in the table, but trailing pressure drops by 50% to 2.5N at 0.866025 max speed.

With the aid of a JavaScript program, we zoom in on the force curve or y axis by a factor of 200 in LS2 below. Even at a magnification of 200, line separation at 0.00 to 0.10 c is unperceivable. At 0.20, a line can be identified easily. For page size sake, we lopped off the tick at 0.999 c from this page. Force at that speed is down to 0.2236N.

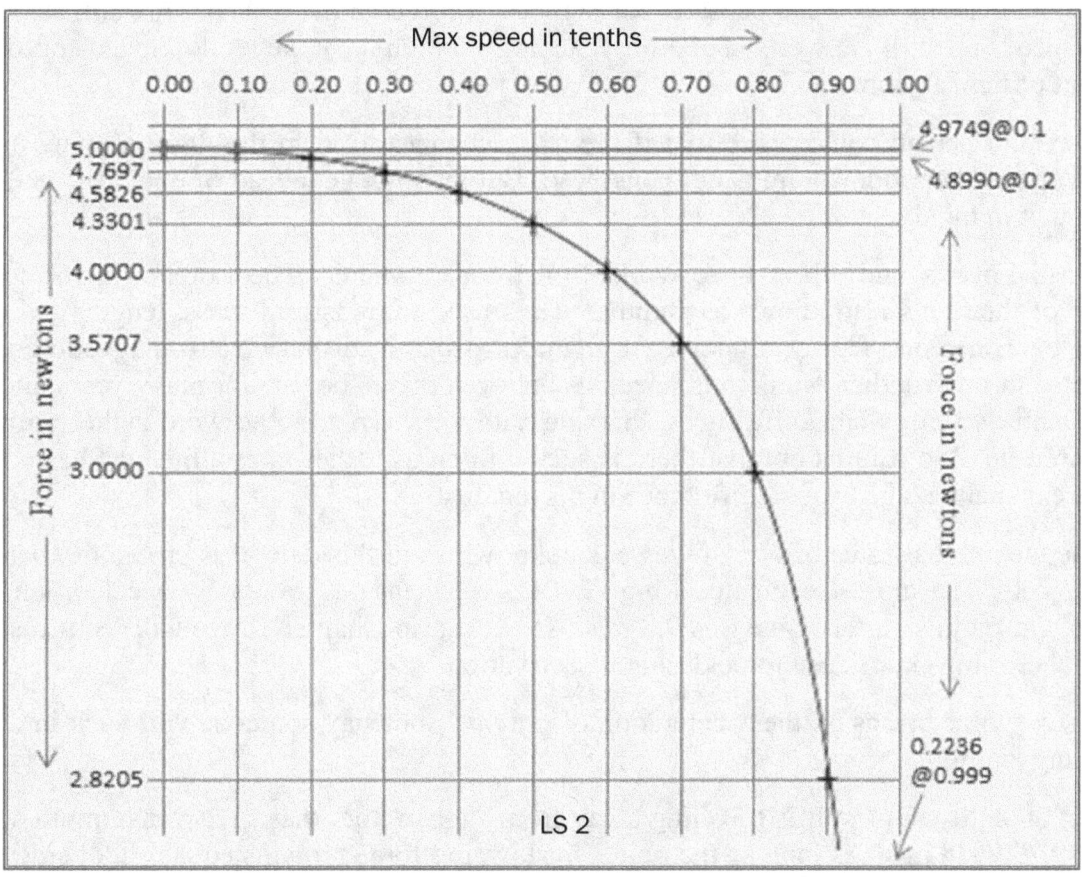

Referring to table VT1 derived from our test, at 600 m/s the device can only exert 4N on one side while the opposing side has at least 5N. However, there may have been an increase of force on that side. At this place in the discussion, that increase may or may not exist. We know there is a net force acting in a negated manner to slow the object. If not dealt with, that opposition will bring the ball to a halt and then reverse it.

But that is not what occurs in our Universe. It makes an adjustment to the object someway to maintain momentum. What are those adjustments?

At slow speeds, the mechanism's force emulates superflux, but something goes awry at high speeds relative to its maximum. How does the universe compensate for a decrease in the trailing force? A lot happens to the local section of the universe when something moves at such speeds. Trailing flux virtually stretches as distance increases; trailing time speeds up as distance increases. Leading flux compresses as distance decreases, and leading time slows as distance decreases. What about mass? Mass increases—the energy portion of the work done during acceleration goes into the object. Newton's first law remains true because the mass increase compensates for the loss of force on the rear; $\rho = mv$ still stands at new speeds up until . . . well, just before reaching c which it can never do. Speed approaches that of light asymptotically.

However, a person could argue that if the mass changes to keep the thing moving at the same speed, then momentum is not conserved. But let's not get ahead of ourselves yet, just keep that in mind.

Hendrik Lorentz and others have worked on these problems. The Lorentz factor solves some of them. It's also known as gamma, γ. It's used to transform mass, length, and time from one reference frame to another. Lorentz also believed that electro-magnetic energy traveled through aether. Numerous scientists believed this to be true for many years into the 20th century. With slight differences, this alternate view replaces the word aether with superforce flux. So it turns out that there is such a thing as aether after all. It just has a little different spelling and a small difference in personality.

While the internal force of our Universe is unknown as of this date, it is brought to bear on all objects. The closest estimate is big G. But, we measure gravity by acceleration, not force. On Earth's surface, one g is 9.8 m/s^2. However, in Chapter 18 we will learn that the formula for big G does use force during its derivation.

The superforce brings on the acceleration of gravity. Someday someone will back into calculating that force.

When an entity moves, it is traveling away from flux vectors that have a maximum speed of 299,792,458 m/s. As long as the object is slow, the force remains equal on all sides. As our test shows, when its speed increases, it tends to run away from those vectors that are assisting to maintain the same speed. To keep the object moving at the same speed when the extra force is removed, the universe compensates for the fall in trailing force by increasing its mass. That we know. We just don't know how. We need to know how it works. How does an increase in mass come about with an increase in speed?

We used the term mass increase because that is the most common conviction; however, keep an open mind. We are going to shovel a lot of dirt on that view in chapter 18.

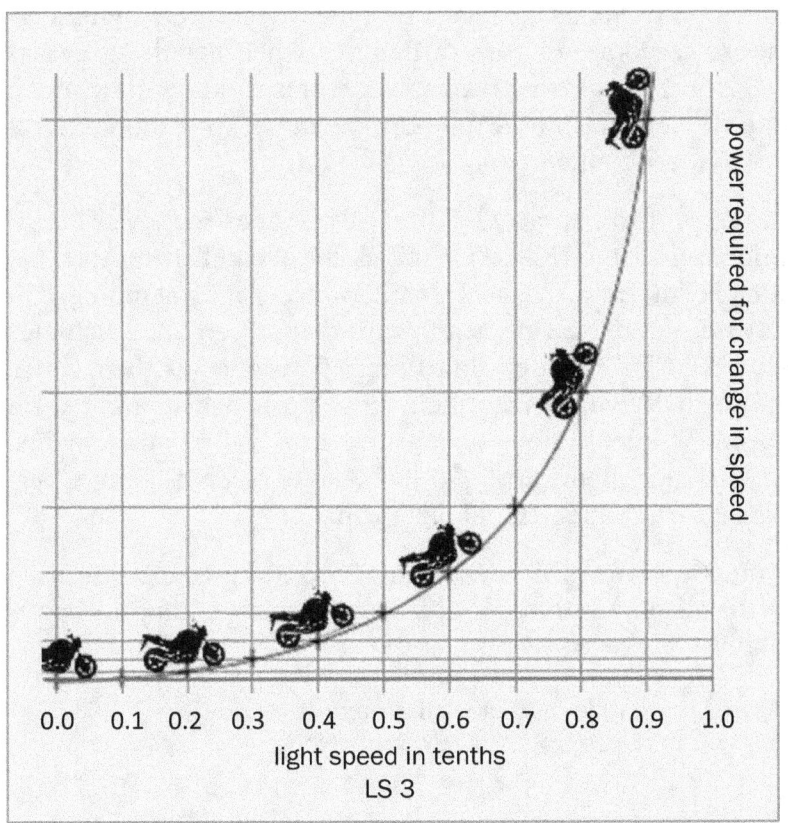

When we turn the drop-off curve on its head, it describes how much energy is necessary to cause a change in speed. The image is still exaggerated 200%, but at this magnification, the increments are more uniform although shown in an exponential manner. Distance between each horizontal line doubles moving from bottom to top.

The type of energy is work. Earlier on, we discovered that no energy is consumed until work is performed. When that work is done over time, it is power. This example uses horsepower to visualize how much extra energy is required to turn a quarter mile consistently in under 10.9 seconds. Each tick on the graph represents horsepower required to run the quarter-mile when that local area is moving through space at the indicated speed.

The tests are performed in a local context, a local point of view, but someone is observing from afar in another reference frame. The foreign observer's position in the universe is unknown to the people having all the fun. Further, the local group has no idea of their actual speed through the universe. As far as they are concerned, they are on Earth on a lonely straight stretch of road between Las Vegas, NV and Death Valley, CA.

Energy should remain the same for all local tests but increase at the observer's position. That's because mass of fuel, air, and everything else has increased with speed unknowingly to locals.

Mass increase is an effect not a cause. The cause is that trailing flux cannot sustain the same force as the leading flux causing unbalanced forces.

Observing graph LS 3 one can imagine how much extra effort is required to overcome the differential pressure opposing any change in speed.

Universe's Mathematics

Being aware of Einstein's concern about folks using relativistic mass when conveying certain ideas, we did a few calculations to come up with difficulties when people use mass alone to express such problems. Suppose some far-off observer wants to know how much horse power is required to make a decent run in a ¼ mile drag contest. And suppose that observer is not very careful about the local context which he is viewing.

The observer runs a computer program using an acceleration of 22 miles/sec^2, and he ignores all friction and aerodynamic drag in the tests. He figures the weight of motorcycle and rider as 500 lbs. That number results in a mass of 15.528 slugs. Understanding that mass increases for every mph increase of the local system toward that of light, he calculates the hp required to run a ¼ mile with an ET (elapsed time) of 10.9 seconds as their world travels through the universe. The locals have no idea their universe is accelerating from zero thru 99.9 percent of light speed. The looky-lou executes the runs in 10% increments, but we post the visuals in LS 3 in 0.20 increments until .90 c. That one receives a more vertical motorcycle instead of one disappearing three feet above the page.

To provide a reference of how much power is needed to overcome the drop in trailing force, the at-rest base is used for distance and time. Notice the tremendous jump in hp required from 0.900 to 0.999 c in Table VT 2 below.

Horsepower required for ¼ mile drag of local groups at various speeds of their universe from zero through 99.9 c.											
0.000	0.100	0.200	0.300	0.400	0.500	0.600	0.700	0.800	0.900	0.999
74.844	75.221	76.387	78.458	81.662	86.423	93.555	104.803	124.740	171.704	1673.980

Table VT 2

When the simulation runs using gamma for all affected parameters, the horsepower required does not change at any local speed through the universe. It is always 74.844 hp even at $0.999c$.

Energy is conserved no matter what speed a reference frame may be moving. That is because our Universe compensates for any change in speed, in any direction, at any time. For every increment a local context advances toward c, all atoms gain in mass. The gain comes from the energy used in increasing that world's speed. Later the term mass will be modified to express that the increase in energy to bring about a change in speed becomes attached to the object as the energy equivalence of mass.

Chapter 15

Information Speed

Say a spherical mothership hundreds of meters in diameter transporting several flying saucers and a crew of hundreds traveling at some outlandish speed passes Neptune. It's headed for Earth from its home base on a planet near Alpha Centauri.

Suppose its trajectory as it enters our solar system is parallel to Neptune's orbit with a speed of $0.866\ c$. Ignoring all the other problems with this example, how does the ship's snapshot appear at this speed to the observers at Caltech through the SpaceX Voyager probe stationed near Neptune?

For that answer, we need to understand what is going on inside the mothership's world. Speed 0 shows the original plan view while stationed at its base. We're going to go through five phases to learn what happens as the spaceship increases in speed.

Just before departure, the pilot tells the navigation computer (NavCom) where to and how fast. When all-is-good sounds, the NavCom engages the engines that are attached directly to the bridge. The mothership with all aboard immediately accelerates to 86.66025% of light speed. This thing is over 2400 meters in diameter, so there is a delay from the time the bridge moves until all other parts of the ship respond. Even though they respond as one unit, information can only travel at light speed.

Universe's Mathematics

The distance between each bulkhead is 300 meters, and the distance from the bridge to the inner bulkhead is also 300 meters; therefore, it takes one microsecond for the first bulkhead to realize it must move. Meanwhile the bridge has moved 150 meters to the right gaining on the inner bulkhead. That is shown in Speed 1. Another μsec passes before the next bulk-

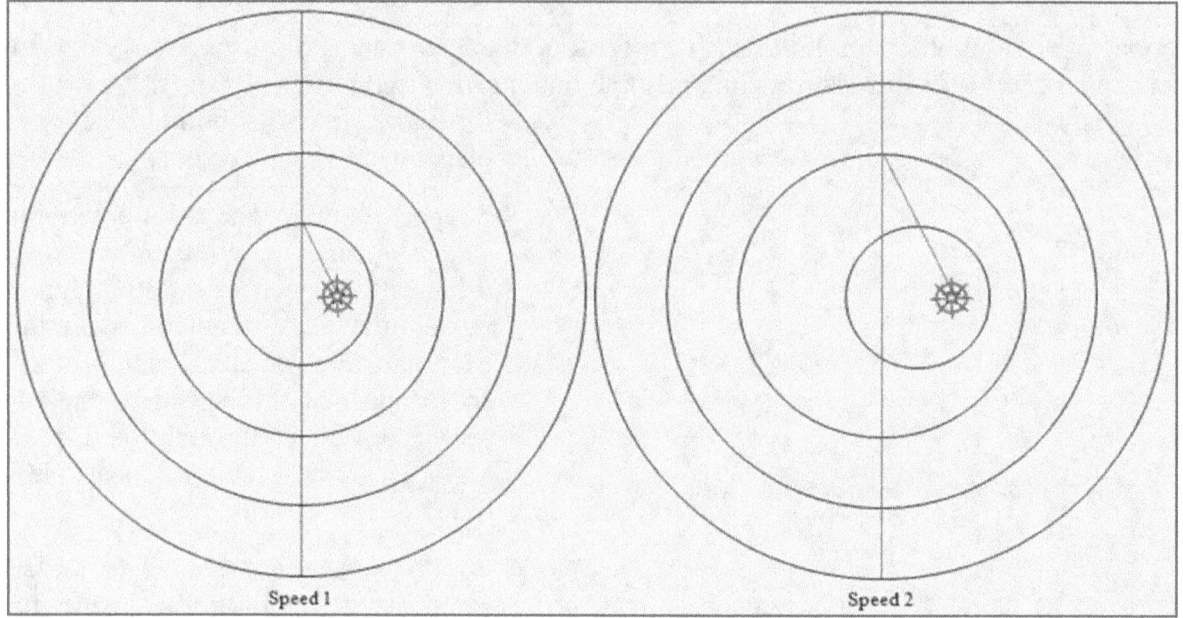

head gets the information, then it too begins to move. Speed 2 shows how the bridge and bulkhead one have closed in on bulkhead two just as it receives the go-ahead.

Three µsec after the initial blastoff, partition three receives the go-ahead as shown in Speed 3. One millionth of a second later, the outer layer of the ship moves forward as in Speed 4.

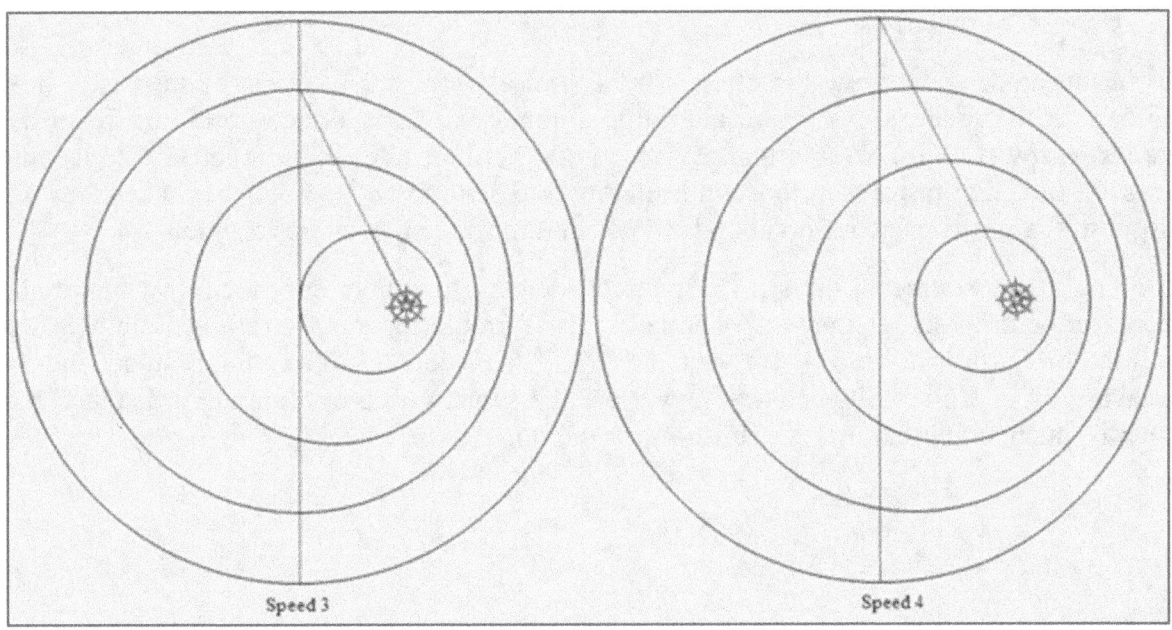

Speed 3   Speed 4

When the ship slows, the change in speed is passed along in the same order. The bridge slows, next the inner bulkhead slows, and then comes bulkhead two followed by three. Finally, the fuselage slows. When the ship comes to a halt, everything returns to the manufactured state as in Speed 0.

Popular belief is a sphere somehow appears as a disk at speeds near c. If this were true, all sorts of nature's laws would be broken: mainly, that every part would receive information far beyond light speed, and those trailing parts would radiate blue light instead of red.

Realize the above conversion is good only when the ship is underway with its own power. If it collides with a large rock or other object, all kinds of distortion takes place under different laws.

One might argue that an object in freefall undergoes another form of acceleration that results in a different shape. That is, the spaceship increases speed from inside out while under its own power. But every molecule of that same ship accelerates at the same time under a differential force. That may also be true. But don't forget the drop-off in trailing force as an object increases in speed. Nature does not provide an opportunity for any object to approach the speed of light under a differential pressure of the superforce.

All atoms closer to the object broadcasting its shadow receive a greater force than those in trail. Humans witnessed the results of such actions when Comet Shoemaker–Levy 9 broke apart as it approached Jupiter July, 1992. Two years later the world watched as its remains crashed into the super planet. This is another example of tidal effect which can pull apart the strongest of materials.

To further understand how this effect works, turn a water faucet on and adjust to a slow stream. At the faucet's exit, there is a large amount of water. Notice how the stream becomes narrower as it grows longer. Even though the molecules cling together, the leading ones are traveling faster than those in trail. And the delta force is stronger because they are closer to the earth's center. Soon bonds break, and the stream becomes droplets.

Looking at the source of the stream, one can detect a parabola concaved downward and then see the droplets separate exponentially. The formula for how far a molecule of water will fall on Earth is distance = $16t^2$ round down. The mechanics of a running faucet and the breaking up of Comet Shoemaker–Levy 9 are the same. Of course, Jupiter's distance formula is much larger than Earth's. It is $41t^2$ round up.

Chapter 16

Space-Time Constant

Previous tests reveal that space-time is a constant; it is always one (1). As length decreases, time period increases. When the period increases, it means that the greater distance between ticks takes longer to execute the tocks—time slows. The time and distance (space) must always be some percentage of the at-rest space-time: multiplying the new length by the new period results is one. This is true for any direction forward or backward or any other angle. Maybe not proof, but hopefully enough for folks to ponder. Using trigonometry, the Lorentz factor can account for an observer at any angle from the object. The next chapter will demonstrate length (base) period (height) as a rectangle making it a little more visible.

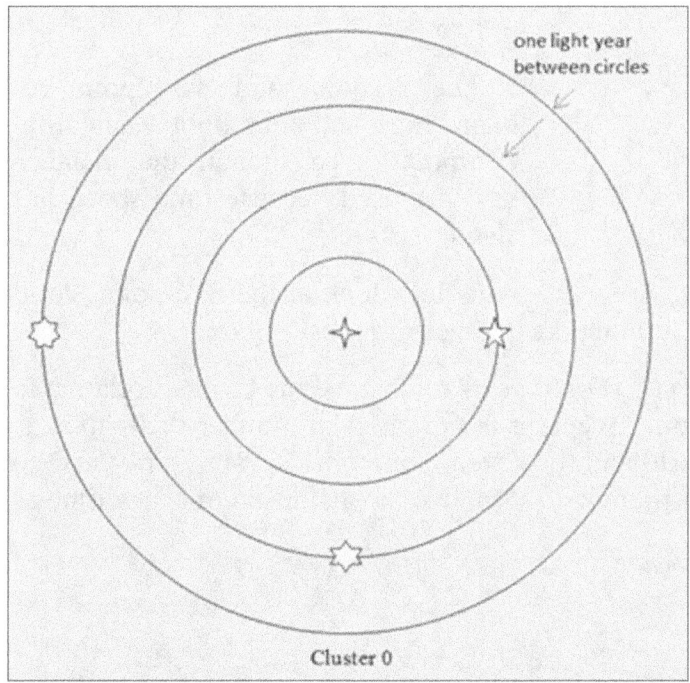

Cluster 0

The last demonstration was to show how superflux reacts when describing a single object. How about how light reacts when several objects are considered?

Before getting to that though, it is necessary to be mindful that superflux only reveals itself through light beams. It's something like air; it only reveals itself through wind and the effects thereof. So this discussion is really how space-time reacts with varying speeds of objects through space.

Cluster 0 contains four stars identified from a four point star through a star with seven points. The four point star is the center of attraction. All names are descriptive as Fourpoint through Sevenpoint. The image represents the stars as observed from a planet orbiting another member of the cluster. That planet is traveling along with the cluster at a speed 0.866025 c. No one on the planet knows how fast they are moving through their galaxy much less how fast the galaxy is moving through space-time.

Be aware that the following text uses both time and time period. When the period is longer, time is slower, and when the period is shorter, time goes by faster. Sometimes it's easy to get them tangled up.

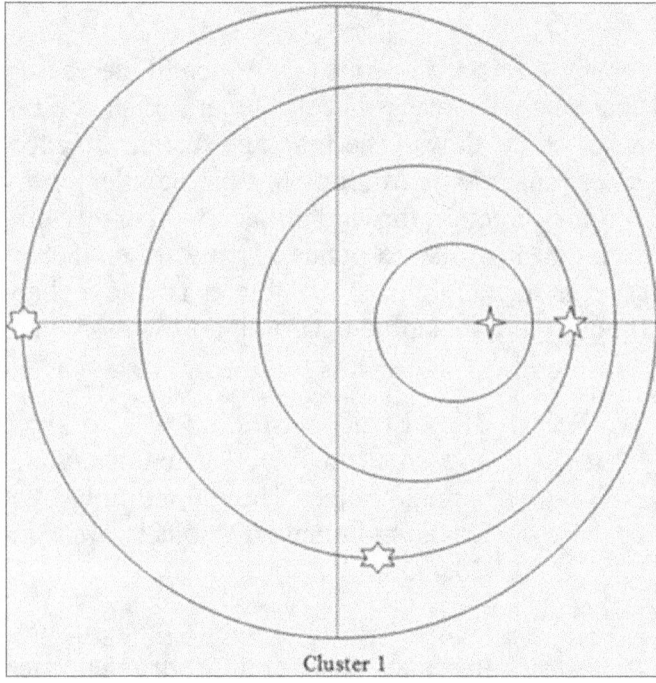
Cluster 1

Cluster 1 describes how distance varies around Fourpoint when referring to it. We've added two additional marks to reveal changes to space-time. Fourpoint is stationed under the vertical line when viewed from its world. But when observed from a neighboring galaxy that remains still, the star and everything in front has decreased in length by 50%. Since space-time is a constant, time has slowed by a factor of two.

Notice Sixpoint and Sevenpoint remain three and four light years from Fourpoint even though the distance has increased because time speed has also increased.

Now let's look at the five-point world in cluster 2 where we added other navigation marks because of repositioning.

There is also an additional dotted circle at 3.6 light years to observe the changes relative to the six-point star since that distance from Fivepoint is not an even number of lightyears. The distance between the five-point world and the seven-point world is six light years, so the image grew larger. We added a right triangle to visualize the distortion of space-time as noted by a separated observer.

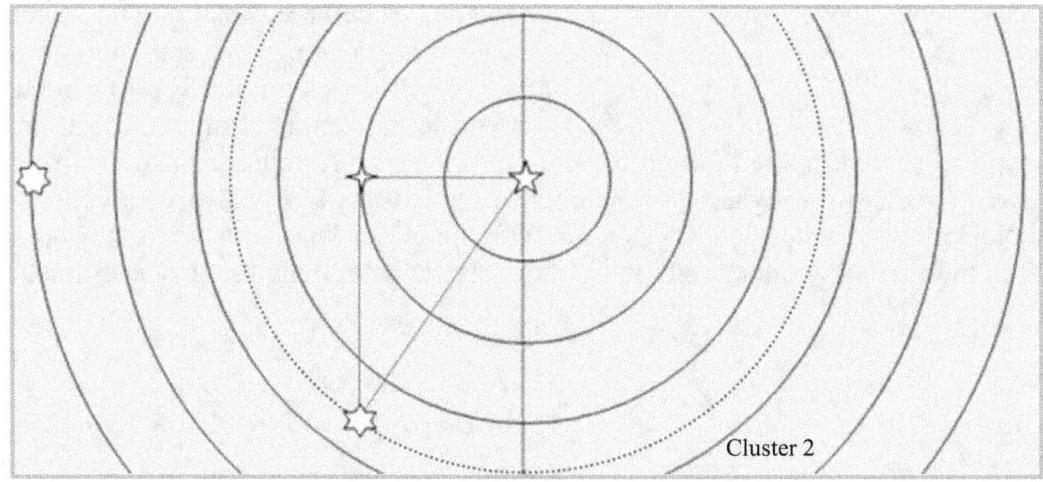
Cluster 2

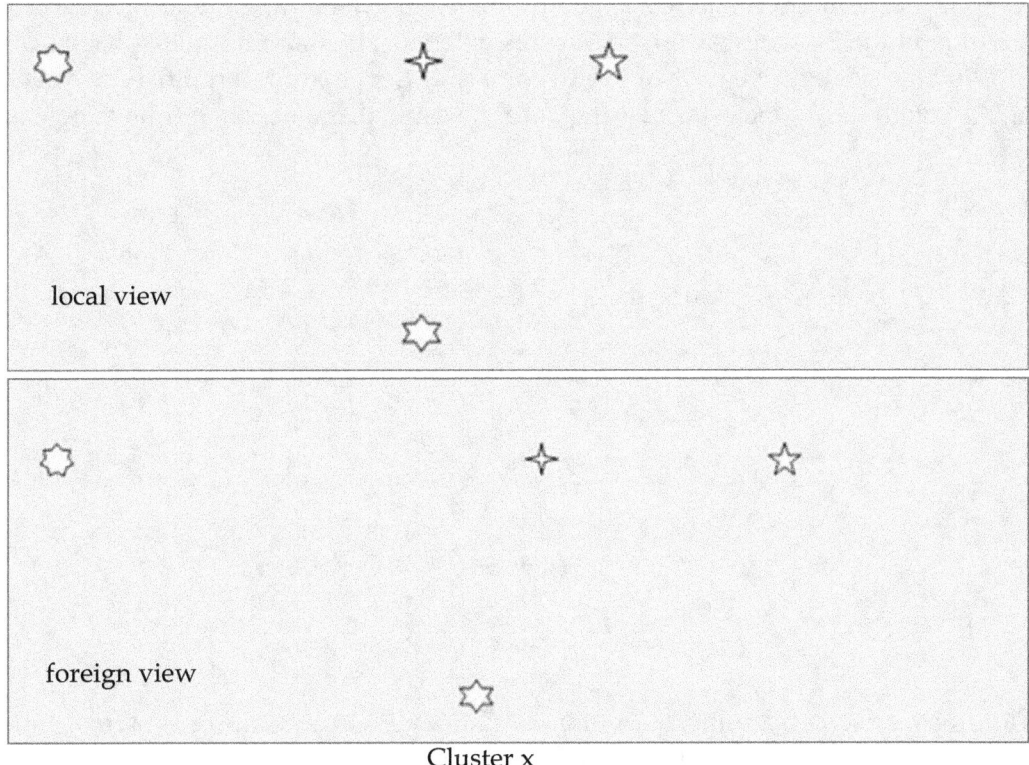

Cluster x

When the same cluster is observed from afar while monitoring the five-point world instead of the four, it appears differently as if from another perspective. To verify what happens, we must turn on space-time markers as in Cluster y.

The five-point has collapsed leading space-time within a large angle and lengthened trailing space-time in the remaining 360° to various extents. The maximum is at its six o'clock. It is still two and six light years from Fourpoint and Sevenpoint and 3.6 light years from Sixpoint. Although they appear farther, they are the same distance when time is considered.

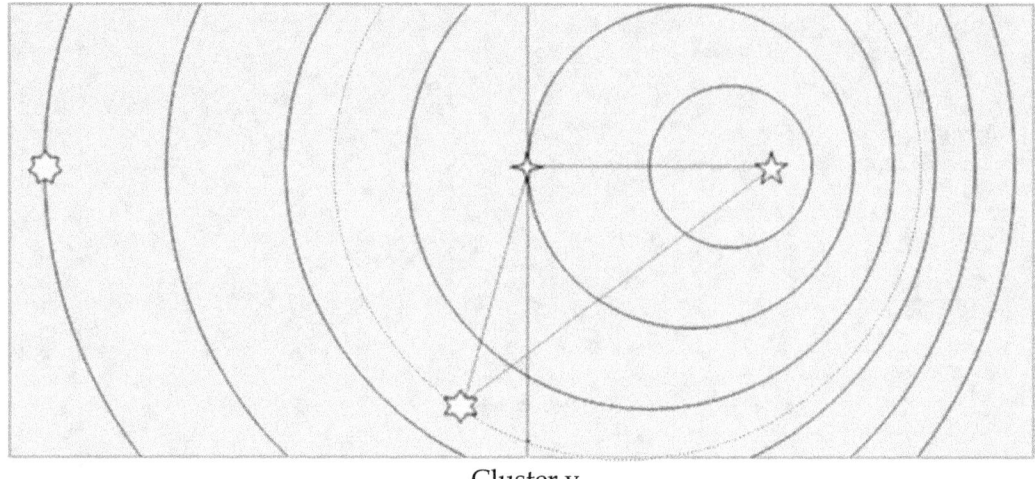

Cluster y

A question may arise concerning the color that appears at the trailing systems. Since the distance grew longer, did the color turn red?

When the focus was on the four-point, the distance to the five-point was compressed, so people on a planet circling the five-point would see a blue shift from Fourpoint. But now that the four-point seems to be receding, it may tend to redshift. How can these variances be reconciled?

If confusion sets in with time period vs time speed beginning on the next page, please refer back to Doppler Effect, Chapter 10, page 67, Tone 2. Studying the graph again may boost your morale.

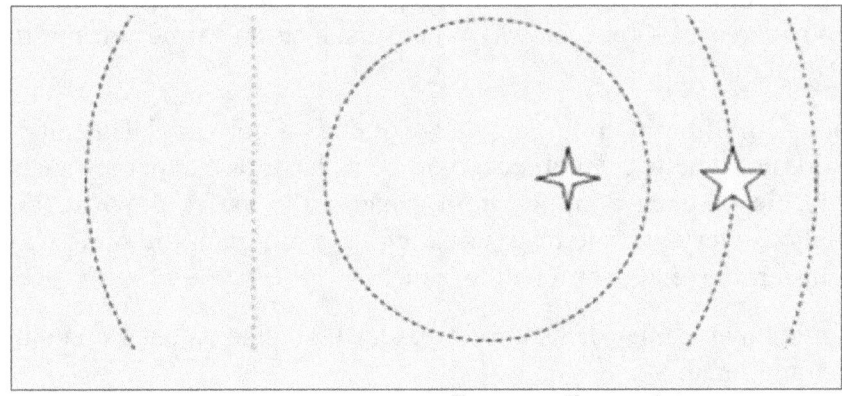
Focus on Fourpoint

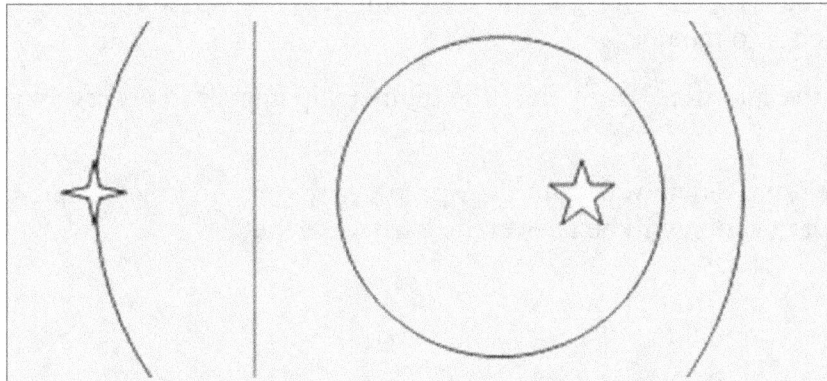
Focus on Fivepoint

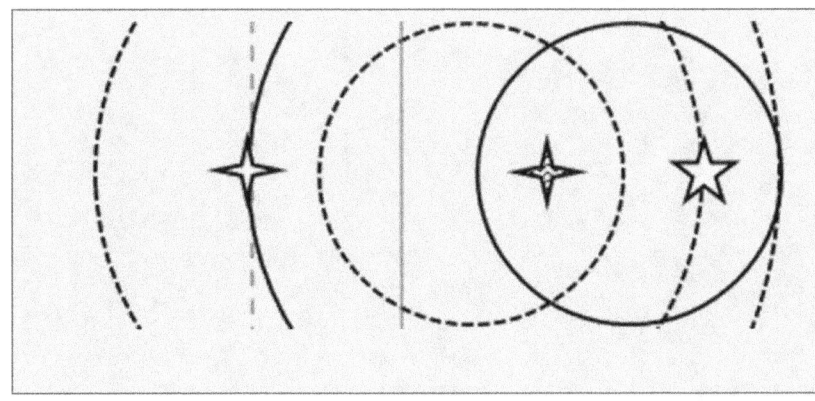
Overlay

When referencing Fourpoint, distance in front is shorter while the time period is longer; trailing distance is longer while time period is shorter. When referencing Fivepoint, the same occurs. The distance is still two light years separating the two. If brain-fog blocks reason, refer back to the local view in Cluster x to restore all positions.

Although the stars are traveling as a group when observed individually, space-time varies for each star. Put a memory sticker on that for later.

Bring on the head scratcher and begin to dig in. When superimposed, the answer to the question of how the mothership appears to observers at Caltech begins to evolve.

In the overlay, both Fivepoints are superimposed. The solid vertical line is the displacement of Fivepoint from its standard position. That is where it is located from a local reference. The dashed vertical line is the displacement of Fourpoint. As Fivepoint extends distance in trail, Fourpoint compresses distance in front. And in front of Fourpoint is Fivepoint, so they cancel each other. Again, Fourpoint shortens every meter lengthened by Fivepoint and by the exact amount. If length increases, time period must decrease; that is, time speeds up. If length decreases, time period increases—time slows. When distance and time are multiplied together, they always equal one.

If space-time varies with each star within a group, at what size does it not vary? This alternate view believes that size does not matter. An object could be as large as a super red giant or as small as a single molecule or even a single atom. Space-time varies down to the smallest of objects. If we analyze every particle of gas making up a red giant the same way we examined our previous cluster, the results will be the same.

Why? Because all matter is at its own center of its universe. Our Universe is just a container for every particle of matter and its universe.

There will be a test on the last two paragraphs, so please commit, "How space-time varies down to the smallest of objects," to memory.

With that in mind, return to the question, "How does the mothership appear to observers at Caltech?"

They observe an oval mother ship as viewed from the SpaceX Voyager. And the scientists applaud because soon another question will be answered. "Are we alone?"

A Sphere is a Sphere

Earlier demonstrations mostly concentrated on objects in trail. To prove that the width also remains faithful requires a different perspective—side-by-side.

Cluster B1 is a composite of stars Fourpoint and Fivepoint when the focus is on each one while the cluster is at rest or at a local view. Fivepoint is a dashed world. Beginning at Fivepoint, the distance to the first circle is one lightyear, and one lightyear separates all other circles. Furthermore, the distance from Fivepoint to Fourpoint is one light year.

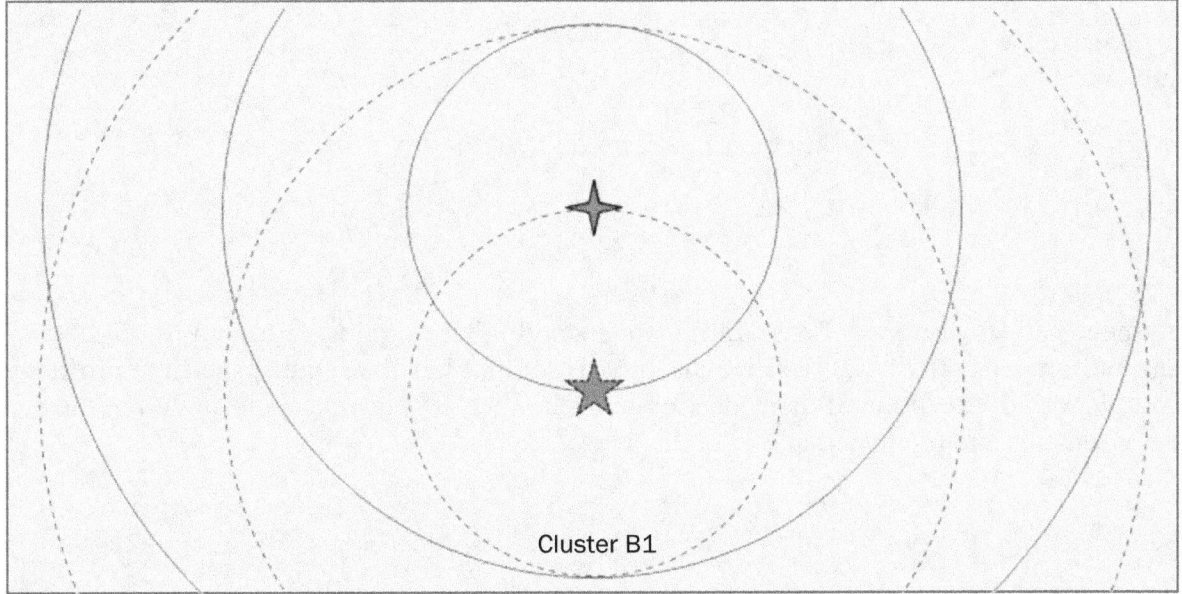

The Fourpoint world is the same, except its circles are solid lines.

The images show only two stars of many in a cluster traveling through their universe at 86.6025% light speed.

Cluster B2 shows how Fourpoint moves in front of Fivepoint when observed from afar. The distance between remains one light year, but its space-time profile has slowed time and reduced length in front. We have no idea what has happened to Fivepoint from this prospective because it is in its own universe—its own space-time. This is true, although it is moving along at the same speed as Fourpoint.

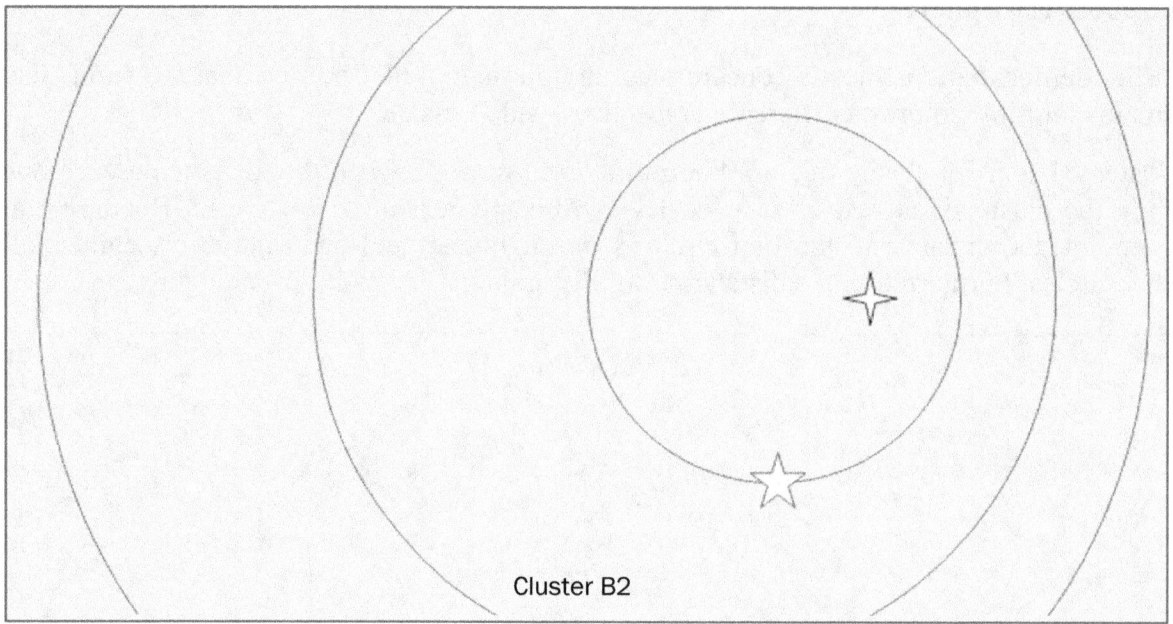

Cluster B2

It's necessary to focus on Fivepoint to understand what is going on around it. When we turn our attention to Cluster B3, we discover Fivepoint has the same space-time profile as Fourpoint with exception of their location to each other. It is leading Fourpoint, but the distance remains one light year apart.

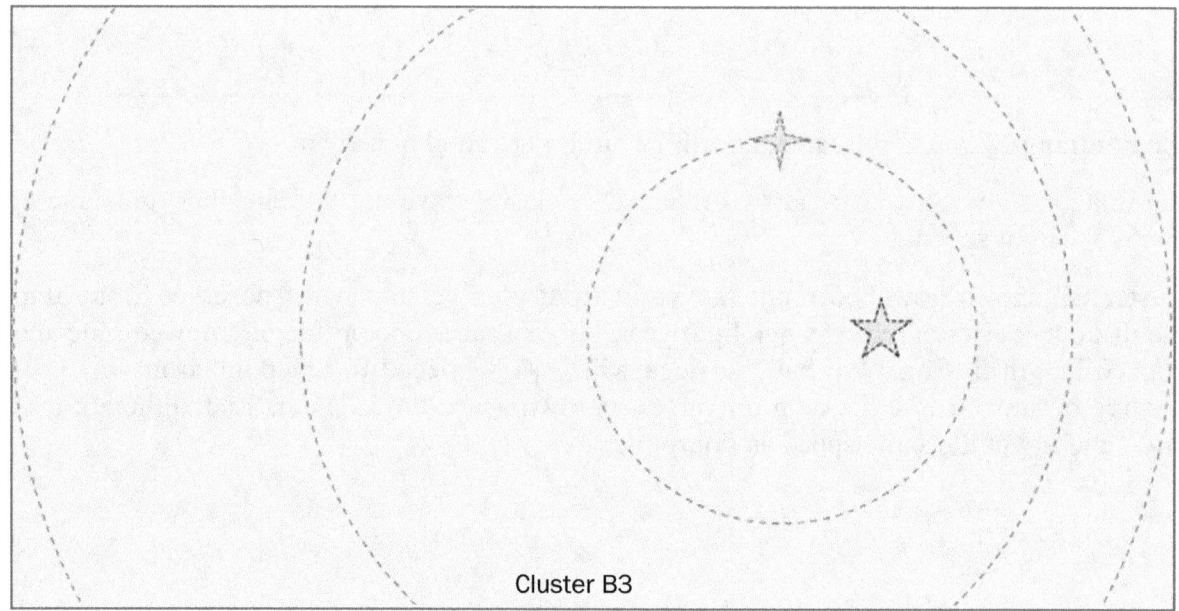

Cluster B3

Every boat creates its own wake as it moves through a body of water regardless of how many are on the lake. And something like the boats, we observe each object's wake as a group moves through space. All things in travel influence the surrounding environment. It could be the train that we used to show how its movement affected sound waves as it traveled forward by shortening the period of a tone in one direction and stretching it in another. Passengers on the train were not aware of any changes because they were part of the steam engine's world. People along the track were removed from its world and could see and feel the effects. People observing Fourpoint from afar can see how it influences its surroundings, and when they observe Fivepoint, they can see how it also affects its space environment.

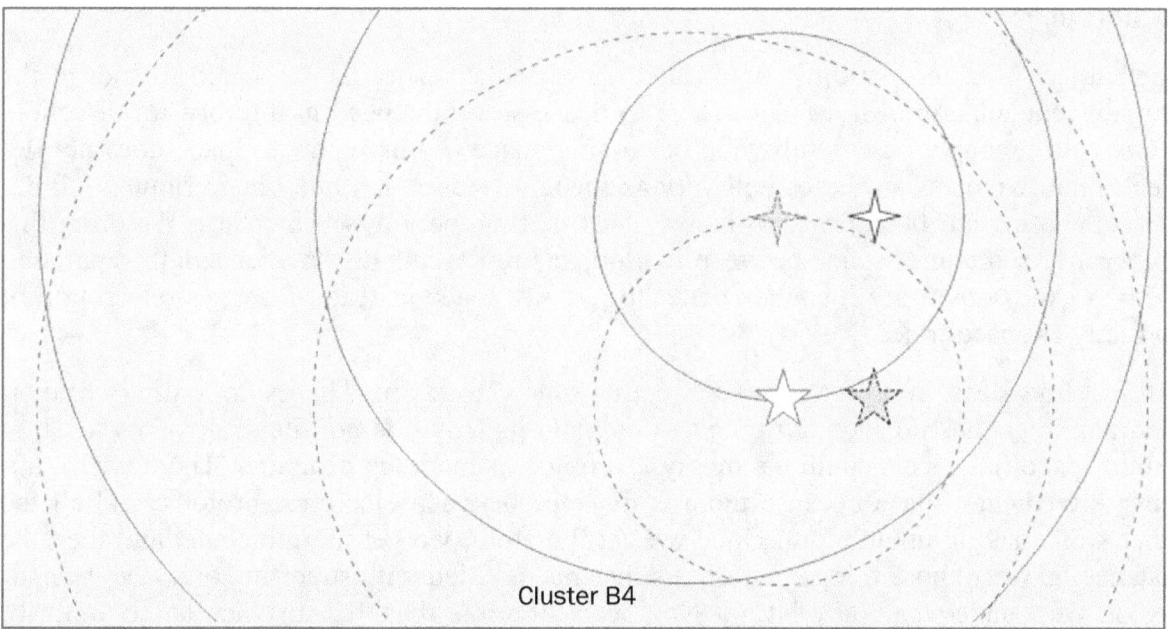

Cluster B4

Cluster B4 is an overlay of both star's effect on space-time, and it reveals how the two remain the same distance apart. As previously mentioned, when one shortens the distance, the other lengthens it, and when one lengthens distance, the other shortens it. There is no distortion. A box remains a box, a sphere remains a sphere, and the cluster remains a cluster.

It's important to understand that the cancellations are only good for the group's configuration as a whole. Time still slows, and length still decreases for each individual unit in the group, but the makeup of each part remains steadfast. The physical mass does not change. It just has much more energy than at rest.

Size does not matter. A galaxy works the same as a small star or a molecule. However, this could raise a question as to whether or not the volume of a box changes with speed. It may seem that one explanation prohibits the other from occurring. We'll have to wait and see.

Chapter 17

Time Dilation

It may be obvious, but very few people give it a thought. There is only one direction for an object to travel either in space or on earth—forward. If it backs up, it's still going forward. If it turns left, it is still going forward. However, observers may attest to its travel from all spherical perspectives. An attribute of forward is some point on any x, y, z axis. Like north or 360 degrees. This author will to defend this statement, however it may be important in the next topic.

The subject of time dilation is predicated on what the reader has accepted as true in the previous explanation of how space (length) decreases with speed in the forward direction. We should probably be careful when discussing space in this manner. Space does not decrease, but two of its attributes will vary as speed increases. Length, one attribute, will decrease; another attribute, time, will slow. That is, time period will increase. We bring this up because sometimes space between two objects represents distance or length separating the two. And sometimes it sounds better to just say space instead of always referring to it correctly as space-time.

But just how does an object's speed affect its time? Heck fire. That is not entirely true either. An object has no such attribute as time, and its length is not the same as its location within space-time. This could get messy. An object is made up of matter. There is also another ingredient—space because there is distance between electrons, protons, and all the other stuff making up an atom. Once we get the atom, we get the molecule, and then the distance between those things. All of this distance is filled with superflux or space depending on your understanding of it all. Now, can all agree that that distance has two attributes—length and time? After all, perhaps an object does have the same two attributes. Of course, chemical actions and nature has much more influence on all material making up things.

Now back to the original question. How does an object's speed affect its time? One rational suggests that when space or length decreases, the frequency of sound increases. That implies that space taken up by that tone is reduced. That length of one cycle is also referred to as its period. If it is reduced to 50% of its original length, there is only ½ the distance necessary for containment of the tone to sound. It has been squeezed, so the frequency must increase. If distance is presented as a percentage of max speed, an idea conveyed for sound can hold true for light as well. That is true even though one is linear and the other exponential.

For example, the distance a cycle period contracts by 50% takes place at ½ max speed for sound. That is 171 m/s. For light, it is different. The speed at which distance contracts by 50% for light is at 86.6025% c, so the resulting speed of an oncoming star and its frequency (color) is more complicated. We won't go there.

Think back a little to when we analyzed the Doppler Effect using a train whistle. The speed was actually a large percentage of mach 1, and the frequency changed quite a bit in relation to the base of 512 Hz. As the train moved toward the barrier, it shortened the distance between the generator and the terminal. It squeezed the space therefore increasing the frequency. Later we will explain the consequences of exceeding that maximum speed.

When an astronomer sees light shifted towards blue, she knows that the star is approaching Earth at a speed which she can calculate. Most everyone concentrates on that shift in frequency. But there is another part of the shift she may include—the star's time dilation. If it is part of a solar system, and intelligent beings are on a planet revolving around it, time is running slower on that planet than Earth's time. Its seconds will vary a little depending on its orbital position relative to Earth, but overall everything runs slower on that planet. The converse is also true. A group of astronomers on that planet could infer that time on Earth is running slower than the time on their planet because they see the universe exactly as earthlings see it. Okay, that just came up, so forget it.

When space is reduced by ½, the frequency doubles. But to locals traveling along with the star the color is the same as if they were at rest even though the group is speeding across the universe at 86.6025% c. The locals detect no change in either, so how does the universe compensate for that decrease in distance and increase in color frequency for those people?

Let's see: suppose we use the space-time constant to hurry the finding along. We know that it is the numeral 1. But while we're at it, we'll add a little more intuitiveness to the constant.

Would it be possible to emulate space-time in 2-D with a rectangle where width represents the length attribute and height represents time? Let's try it.

Universe's Mathematics

If an area of a rectangle stands in for the constant, the product of L·W = 1 should always result in the same area. In this case 1 because the rectangle in the image below is 1 m wide by 1 m tall. If the width changes a percentage of its original length, the height should change proportionally.

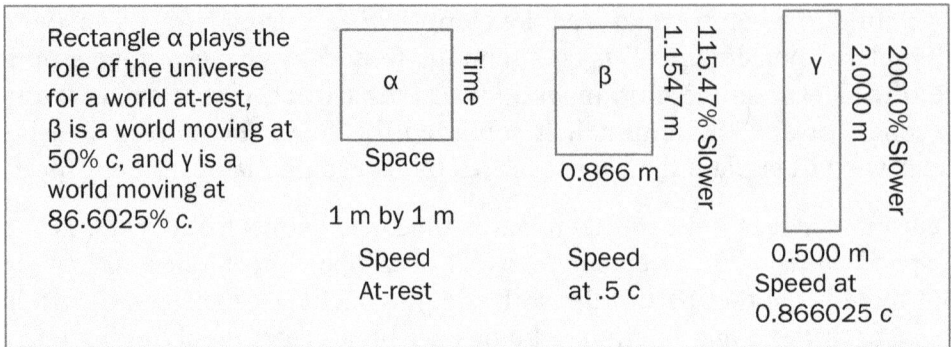

Rectangle α plays the role of the universe for a world at-rest, β is a world moving at 50% c, and ɣ is a world moving at 86.6025% c.

The Lorentz transformation formula is $\frac{1}{\sqrt{1-\frac{v^2}{c^2}}}$ also known as gamma (γ) where v is the moving object's velocity, and c you already know.

It may be used to compute relativistic length: $L_0 = L\sqrt{1-\frac{v^2}{c^2}} = \frac{L}{\gamma}$

where everything right of the equal sign is moving; everything on the left is not. L_0 is the observed length, and L is the measured frame length (local length). The subscripts are probably different than most readers recognize. They have been identified to the author's understanding.

We can cheat a little by just setting $\frac{v^2}{c^2}$ to a percentage of light speed, and go on from there.

At rest, the ratio of proper time period to length is 1:1. Using gamma at 50% light speed we get a new width of 0.866 m. Now divide the constant by the new width $\frac{1}{0.866} = 1.1547$ m.

Since height also means time, what was once one second has become a 1.547 time period. Time passes slower; it now requires 154.7% more time to move across that new period.

Moving on to ɣ, at 86.6025% c, length is 0.500 m. Again dividing 1 by 0.5 gives the new side 2 m with a time period of double whatever it was at rest.

Another way of looking at time is through frequency.

For this purpose, we will look at tone instead of color. What can we do to inhibit that tone increase for the local context? Since the length goes from 100% to 50% at 86.6025% c, the intermediate period follows the length, and it is also reduced by the same amount.

Inspection of the following graphic tells us a little bit of what's going on.

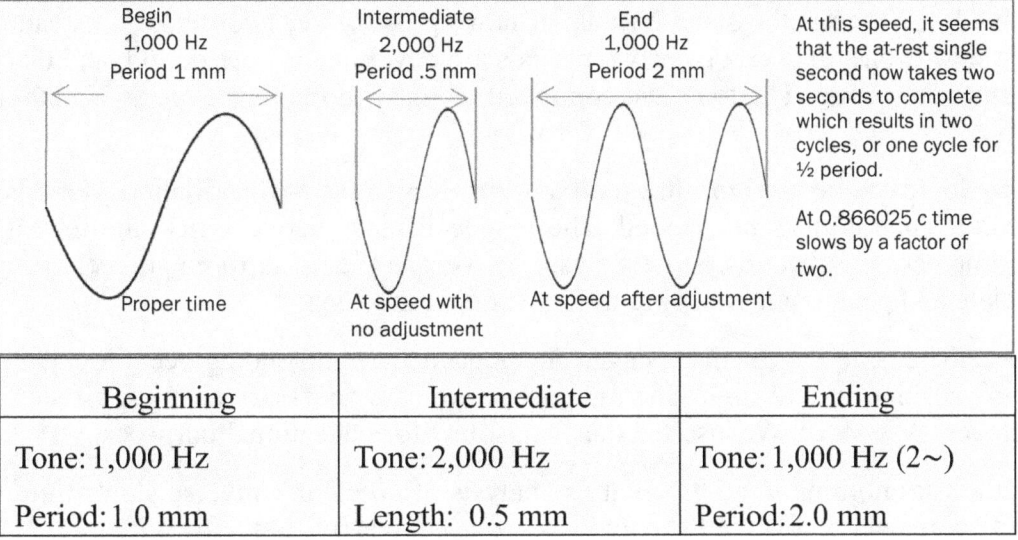

Beginning	Intermediate	Ending
Tone: 1,000 Hz	Tone: 2,000 Hz	Tone: 1,000 Hz (2~)
Period: 1.0 mm	Length: 0.5 mm	Period: 2.0 mm

The procedure the universe uses to slow time is shown in three stages. We must have play time again to understand what's going on, so ring the recess bell.

A solar system is going around its galaxy fine, and then, unknown to local inhabitants, a giant force increases its speed. Suddenly, astronomers from far away notice the star's increasing speed. Using their upgraded equipment, they zoom in and watch and listen to the change in frequency of a single tone. A lady takes charge, and her cataloged history of one kHz of the tone is known well to this group. She can hear the change as it increases slowly from 1,000 Hz. The tone finally settles at 2,000 Hz. Betty, that's her name, uses gamma to determine the systems space-time length. She is not surprised that it has decreased by 50% since she clocked the system in at 86.6025% c.

Then it came to her like a bolt of lightning that the decrease in space for the tone to sound forced it to double. "Umm, that's just like the Doppler Effect," she told herself. Then she wondered what the inhabitants heard. Did they hear a change? If not, why not?

Since the new length is in percent, the new frequencies must also be kept in percentage. Then it occurred to her to determine the ratio of the at-rest frequency to the new length. It was 1.0:0.5. She already knew that time had slowed by a factor two, but she wondered why. Then she slapped her forehead with her open palm. If we keep the same ratio between the frequency and the new length, it should work out fine. So in her head, she placed the 1 over the .5 and the result was 2. Okay, two mm is the period of the new tone, but how does the 2 kHz fit? Another, "Umm." Later, she knew she was on to something. She kept right on thinking. What about tone? What do the locals hear?

Then she realized that since the period of the tone doubled after correction and the frequency doubled, the tone is the same because it now spreads over two periods instead of one. The two new cycles of 2 kHz over two periods are now one kHz per period, and that period represents one second. The tone just continued along with the new time as if nothing happened.

The same logic can be used for time, and that's the objective—time dilation. Hertz is cycles per second. Time is ticks per second. The longer distance between tick and tock means a longer time between seconds. Time slows. She went to the break room and celebrated with a chocolate and peanut butter cookie from a vending machine.

However, there is no intermediate stage. The intermediate and ending occur together. It's an ongoing transition in real time. The universe makes the necessary changes as acceleration of an object takes place. We inserted that step solely for educational purposes.

This author has nothing to add to Betty's analysis of *how* our Universe slows time, except maybe a comment on how the Lorentz Factor is used to arrive at relativistic time.

$T_0 = \dfrac{T}{\sqrt{1-\dfrac{v^2}{c^2}}} = T\gamma$ where T_0 is observed time; T is local time, and you know the others from earlier presentations.

Use the relativistic equation to calculate the time change after understanding how it takes place.

Chapter 18

Flaws of Relativistic Mass

Warning: This chapter is very negative on the use of mass and its relation to the speed of light. It may be frustrating to some readers, so don't feel obligated to read it. It is difficult to refute any accepted theory, but that is probably what this subject attempts. In fact, it is so negative that you may skip it and go to chapter 19, but if you want to find out some of the problems with relativistic mass, please read on.

This topic is to demonstrate problems with a change of mass with an object's speed. It seems that this author in not the only person having such difficulties, and since this book is about how's and why's, we hope to bring to front why relativistic mass leads to problems.

Rod Nave of Georgia State University had this to say on their web-site. Actually, this author doesn't know if it is his words or not, but the following was copied from http://hyperphysics.phy-astr.gsu.edu/hbase/hph.html.

> Even though circumstances like that described at the Cambridge accelerator are conveniently described by assuming an increasing mass, that is not the only way to describe these experiments, and there are problems with the concept of variable relativistic mass. Einstein's point of view is described in the following quote:
>
> "It is not good to introduce the concept of the mass of a moving body for which no clear definition can be given. It is better to introduce no other mass concept than the 'rest mass' m. Instead of introducing M it is better to mention the expression for the momentum and energy of a body in motion."
>
> $$M_? = \frac{m}{\sqrt{1-\frac{v^2}{c^2}}}$$

Since mass is really an effect of gravity, this alternate view believes that mass doesn't literally change with speed. However, virtually it changes through previously stated actions of the universe. Mass change is used mostly to explain why nothing can exceed the speed of light. Sometimes the term mass should not be used at all when exploring c. According to Einstein, it is less confusing to use energy equivalency. Who's going to argue against Einstein? Perhaps this text will aid in his argument against using relativistic mass at all.

After saying that, we give in once again to tradition and refer to the amount of material making up an object as its mass. How does the universe create an object's additional mass as its speed increases? The accepted answer is that the energy used to accelerate the object goes into it as mass. This may be true. But how? What does the universe do to affect this change?

Universe's Mathematics

The following is another of those EWAG's.

A previous discussion led to the conclusion that a sphere is really a sphere regardless of speed. Again, the speed of the object for this rational is taken to be 0.866025 *c*. A length of 50% reduction makes things a bit easier to understand. Earlier demonstrations implied that the width is also reduced by 50%. Accepting this to be true is a big ask, and it is necessary for your willing suspension of disbelief, only for a moment though.

Taking an orb and reducing it to ½ its original size must also change its density, and that we haven't discussed. Density is the orb's mass to volume ratio, so when an orb's radius shrinks by ½, its volume decreases eightfold.

The equation for density is $D = \frac{Mass}{Volume}$ or $\frac{M}{V}$. V is not velocity. To solve for the mass part, it is just $M = D * V$.

Let's say the at-rest speed of a globe's mass is 2 kg; its radius is 2 m. When a detached observer measures this globe's radius at 86.6025% *c*, it will be 1 m. Remembering the formula for the volume of a sphere, it goes from

$$\frac{4}{3} * 3.1416 * 2^3 = 33.5103 \text{ m}^3 \text{ to } \frac{4}{3} * 3.1416 * 1^3 = 4.1888 \text{ m}^3.$$

This is where it is advantageous to use *Alternate View's* interpretation of mass as a value which represents the number of atoms or molecules within its body. That number cannot change regardless of where the body is located or how fast it is moving. Mass also remains true wherever it is located, but it is not true to a foreign observer when its speed is near that of light. The number of atoms within an object remains the same throughout its history. Fast, slow, near a black hole, or free floating in outer stellar space, it just doesn't matter. Oh, there may be a few electrons temporarily abandoning the object at times, but they will return sooner or later, or the object itself will transmogrify into something else. But the number of particles remains the same most of the time.

Using the number of particles within an object instead of its acceleration by a force, we can continue describing how mass changes with respect to speed. The important part is calculated above, the change in volume.

Let's give the orb a name. Devon. Perhaps Devon has a total of 167 moles of *x* within its body. It remains true for all viewers at any speed. What is the density at speed 1 where the amount of material is 167 mols? From this prospective, it is the same as having a mass of 2 kg, so we substitute 167 mols in place of mass.

At speed 1: $\qquad D = \dfrac{167 \text{ mols}}{33.5103 = \text{m}^3} = 4.9835 \text{ mols} / \text{m}^3$.

At speed 0.866025 c: $\quad D = \dfrac{167 \text{ mols}}{4.1888 \text{ m}^3} = 39.8682 \text{ mols} / \text{m}^3$.

Density has increased eightfold. The change in density tells us that something else has also changed dramatically. The driver of the increase is Devon's smaller radius. This we know. But, what other transformations take place?

Did the decrease of distance only affect the volume? That question is one concern; there are others. What else happens when something shrinks by 50%? Since space-time is a constant, could the mass have grown because the distance shortened? It's possible. When something gets cut in half, something else doubles. That something else that doubled could be its mass. Shortly it will be shown that when mass is computed relative to c, the same formula and the same process used to derive time is also used to arrive at the mass value, so it must double just like time doubles at 86.6025% c. These things are only thoughts for now, and rather difficult to get our heads around—just other things to ponder.

After gulping all that down, let's determine how the mass changed with the new method.

Mass = D * V. New density is 39.8692 mols / m³. New volume is 4.1888 m³.

Mass = 39.8692 * 4.1888 = 167 mols which was substituted for the original mass of 2 kg. So, the rest mass did not change at all, but it is supposed to according to relativity. Something must have gone astray. Furthermore, what happened to all that extra energy required in getting Devon to 86.6025% light speed?

It became attached to all particles making up Devon, but it did not change to matter. It followed the law $E = mc^2$ where some quantity of energy is equal to some quantity of matter. That means the universe instills the energy into every atom making up Devon, but it remains energy not matter. There is no physical change within the makeup of the object.

Gravitational Problems

Mass was supposed to change, but it did not when presented as a relation to density. Calculating the mass using gamma says it did. What's going on?

Why does mass cause such a problem when it is moving near light speed? Because it is an effect. Its relation to something else gives it measure and gives it some cause—gravity. And there is no mention of g anywhere in any formula for calculating an object's relativistic speeds. All causes of the effect are changed, so the effect must also change. Could there off-setting properties?

Universe's Mathematics

To what cause do we owe the effect of mass? $\frac{\text{force}}{\text{acceleration}}$ = mass. We need to know where the force comes from, and we need to understand where the acceleration comes from. It is Big G.

We're going over this very carefully because its understanding is what must come before the term mass is appreciated, and why big G leads to such difficulties when trying to understand our Universe, and why this alternate view of the universe prefers to use flux transparency index instead.

Gravity is defined as $G = \dfrac{FL^2}{M^2} = \dfrac{L^3}{MT^2}$

Just what the heck does all that mean? Well, we have to explain the identifiers before understanding it, the first of which is G. That means the universal gravitational constant. The next is F which means Force in Newtons; then comes L^2 for area (actually, L by W). Underneath that is M for mass, and it's squared. When those are divided, they are equal to the final formula with the additional symbol of T, for time, which is also squared.

You probably want to know how FL^2 becomes L^3 and how M^2 becomes MT^2.

Applying those terms with SI units (International System of Units) we get

$$N\ m^2\ kg^{-2} = m^3\ kg^{-1}\ s^{-2}$$

where the negative exponent means it goes under the numerator. It's the denominator.

To most folks it appears as

$$\frac{N \cdot m^2}{kg^2} = \frac{m^3}{kg \cdot s^2}$$ where m is meter and not mass.

M is replaced by kg, and T is replaced by s. The middot or interpoint means to multiply. A space or non-space is also used as a multiplication sign.

Still, how does the left hand side become the right hand side?

Okay, 1 N = 1 kg·m/s^2, where 1 N of force will accelerate 1 kg of mass 1 meter per second per second or 1 m/s^2. So 1 kg·m/s^2 ($ma = F$) replaces the force, N, above. Let's do it in steps.

$$\frac{\cancel{kg} \cdot m \cdot m^2}{kg^{\cancel{1}} \cdot s^2} = \frac{m^3}{kg \cdot s^2}$$

After many years with varying results, people that do such things have arrived at the 2014 value of G. It is

$$G = \frac{6.67408 \cdot 10^{-11} \cdot m^3}{kg \cdot s^2}$$

116

Some readers may notice something peculiar about part of the Big G equation.

(SV) Specific Volume $= \frac{V}{M} = \frac{m^3}{kg}$ where V is volume; big M is mass and little m is meter.

Is specific volume part of the universal Big G? If it is, it creates a heck of a problem because at speed, the volume is reduced by some value and time slows by some value. Taking our example with a decrease in length of 50%, the volume is reduced to 12.5% of its original capacity. At the same time, seconds slow by a factor of 2.

It is just another thing that screws up mass as it varies with speed. Since big G has a relation to volume, and mass is an effect of such, does gravity change along with an object's speed through the universe?

The volume of a cubic meter decreases by a factor of eight when moving at .866025 c as our volume did when calculating density above. If that is true, then gravity has to change by some factor also, and if g changes by the same relation, what happens to everything else? Maybe some of the changes cancel producing no change in big G.

Warning: Trying to figure out things like this is what dried up this author's hair follicles, so don't dwell on it too long.

There is another problem regarding how relativistic mass is computed. The equation for that is shown to assist in why it creates such a problem.

Gamma is used to compute relativistic mass: $m = \dfrac{m_0}{\sqrt{1-\frac{v^2}{c^2}}} = m_0 \gamma$

where m is the proper mass and m_0 is local mass. Notice it has the same equation as time in the previous topic. And therein lies a problem. Mass has a physical size: length, width and other properties, but the equation considers none of them. If the distance between atoms diminishes with speed, the object would soon become so small it would no longer exist. However, earlier we showed that the object doesn't shrink when individual particles are analyzed, so something is wrong.

In the previous topic, gamma was used to compute relativistic length and time. Since only time of an object is considered when calculating its mass at some speed, it has to result in errors. Both size and time should be considered if physical properties change with a change in length. When both space-time properties are accounted for, they offset each other, and no mass change comes about. It goes back to that space-time constant thingy.

We must admit that this whole mass topic was not written to better understand mass, but to better understand problems when it is used as something it is not.

Transparency Index vs Mass

Assuming you committed, "How space-time varies down to the smallest of objects," to memory for a test. This is it.

Big G affects objects by way of volume as per the formula $\frac{L^3}{M \cdot T^2}$.

We can visualize a relationship between the fraction L^3 over M and transparency index, the larger the ratio between an object's volume and its quantity of matter, the smaller the index. If matter is crammed inside a smaller area, the shadow will be darker, and it will create a greater differential force causing a larger acceleration between two objects. If there is less matter within that same extent, the differential will be less. The same is true whether the force is negated as in gravity, or positive as in superforce.

Does the universe even care about the distance between subparts of an object? Each individual atom is in its own world as explained earlier. Every one of them is in its own space-time. When time slows for the atom in front, it speeds up for the one in back. When distance decreases for the one in front, it lengthens for the one in back. It is a recursive act. Every particle of matter inherited identical characteristics at birth. Again, all individual particles' effects of space-time cancel, so the unit as a whole obeys all laws, and the length and time properties of each one's universe remain in order.

When substituting flux index of objects at a great speed, all properties remain the same; volume decreases in the same proportion as matter.

Now let's revert back to what Einstein said. Translation—don't use mass in a relativistic setting.

The real explanation of why nothing can reach the speed of light is that its trailing force drops to nothing at c. There is less force remaining to assist in any additional increase in speed. The mass increase is only virtual—it doesn't happen. The extra energy required to account for every increase in speed remains attached to the at-rest mass as energy equivalency.

Having all this trouble with relativistic mass brings up a question: is it proper to use the Lorentz Factor out of context? By context, we mean taking it away from space-time, which is not matter, and attempting to use it to measure something that is matter. Matter does not have, nor any equivalency of, space-time. The closest the two have in common is energy and embedded distance between atoms. Matter came from super energy, and space is the part of super energy that did not become matter. Here we do not mean vacuum. We mean space-time separates the atom's orbitals, not a vacuum.

Chapter 19

Expansion

Expansion is a born property of the universe. As discussed earlier, an explosion may very well have brought about composite energy, but an explosion did not bring about our Universe. Expanding composite energy gave it birth. Everything that came about during that cooling off period is still in motion today. The most reliable clue of expansion is the redshift of distant galaxies and quasars. While this is a clue for expansion, larger redshifts of more distant objects lead to a conclusion that the expansion is accelerating. That is a conclusion that may need challenging. Also, care must be taken that the receding edge is not the same as expansion. One is at c; the other is at an unknown rate.

When referring to distance in this context, the fact that they are all increasing only means that they are serially cumulative while moving at the same speed. That is, any object two times the distance from another is moving away at two times the speed. If the distance doubles again, its speed doubles again. This is a mathematic truism. Please don't confuse expansion distance with a certain span covered by some object traveling at a given speed.

Expansion of any whole acts over a distance at some speed, it could be as large as the universe moving at a great speed or as small as a short, skinny piece of steel moving at a slow speed—movement slow enough to be measured in nanometers per second. When a poker gets red hot, the molecules expand at the same rate, but those near the tip move quicker relative to those near the center. Each molecule is connected in series in every direction. The steel will become fatter and longer because there are more molecules lengthwise than edgewise. Since each is expanding at the same rate, and they are connected, the end must move at a rate equal to the sum of all in between. Most engineering handbooks will tell you at what temperature and at what rate and how far the poker will grow. It's just a matter of size, poker vs universe. But there is no handbook for our Universe.

Confusing things even further is the fact that galaxies within local groups approach each other: the Milky Way and Andromeda for example. They are on a collision course, so where is the expansion part taking place? Perhaps it's like watching the mighty Mississippi flow toward New Orleans. Currents near the banks are free to flow upstream, swirl about in any direction, but in the end, all the water ends up in the Gulf. Like the currents near the banks of the mighty river, large groups of galaxies follow their own courses, but expansion provides separation between those large groups of matter. And the severance is in every direction and at the same rate. Besides, our Universe is pressurized. It must expand like anything else under pressure until a wall or something else restricts it.

Recede 1 and 2 depict a slice of a spherical view of what the universe enlarges into while earlier graphs show a lateral view. Recede 1 demonstrates how your Universe grew from 1.9 BYO to 4.9 BYO as it replaced super energy.

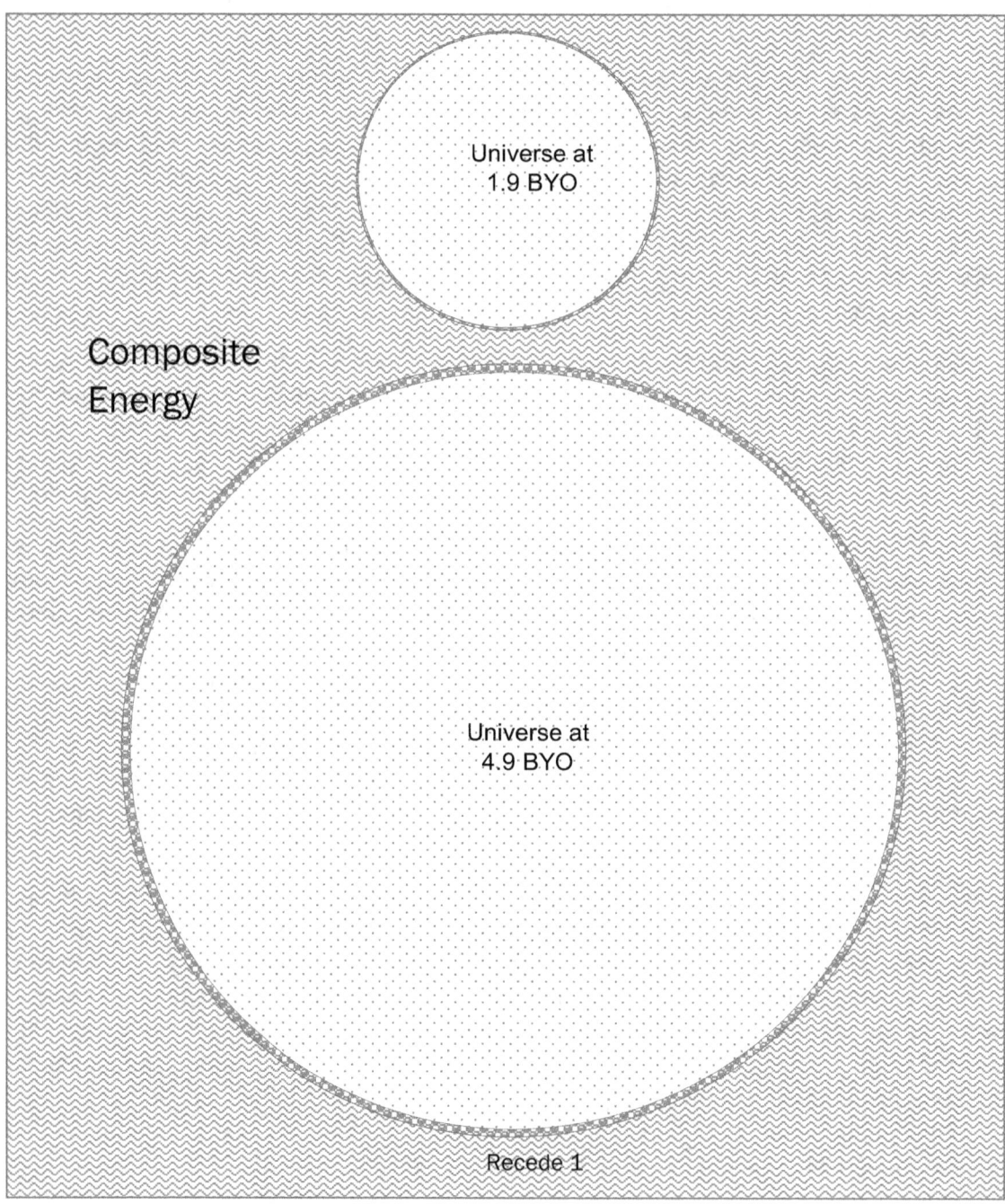

Recede 1

Mine did the same but the separation is too small to see the difference.

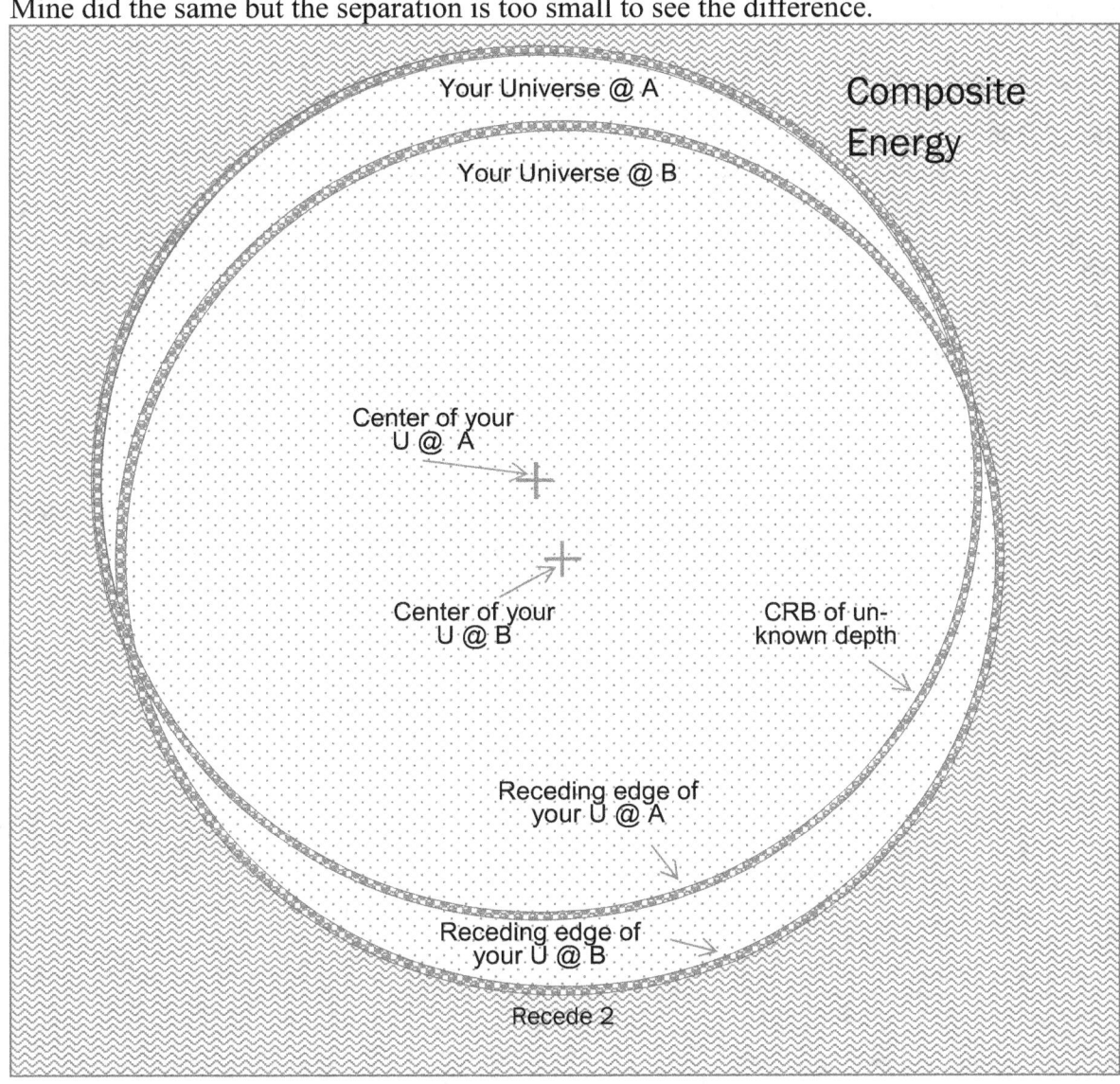

Recede 2 demonstrates how the center of your Universe travels along with you. You were born at position A and then somehow moved to position B after growing up. During that same period, the edge of your Universe receded into the composite energy field at 300,000,000 m/sec. Of course this could smear the difference between receding edge and expansion where expansion includes all of composite energy and all universes. Looking back in time through the CRB to composite energy, tells us this.

But what is the other concept of expansion? To answer that, we must take into consideration the growth of composite energy that brought about space and matter in addition to our current Universe and all others making up the entire collection. This author has no name for these objects when taken as a whole because, generally speaking, the universe means everything that exists.

The closest we can come, considering our present understanding, with that description is to take a loaf of raisin bread and bake it. Everything within the loaf expands: dough, tiny air bubbles, raisins, and perhaps yeast. Say the raisins represent universes, and that they swell at the same rate everything else does, we can see that the raisins don't really expand into anything at all. Everything is just growing. Of course, that leaves us to worry about the crust of the loaf, but there is a limit to our imagination.

To restate both of the above:

- Space is that part of composite energy that continues to expand as it was before the other three forces became matter.

- The edge of our Universe moves into the area as Grand Energy generates more space and matter. Our Universe, along with all others, continue growing at c.

Bungee Experiment

This experiment requires a four meter (or foot) bungee cord and five clothes pins.

Take a four meter bungee and clip five clothes pins one meter apart beginning at its left end. Mark the pins *a* through *e*

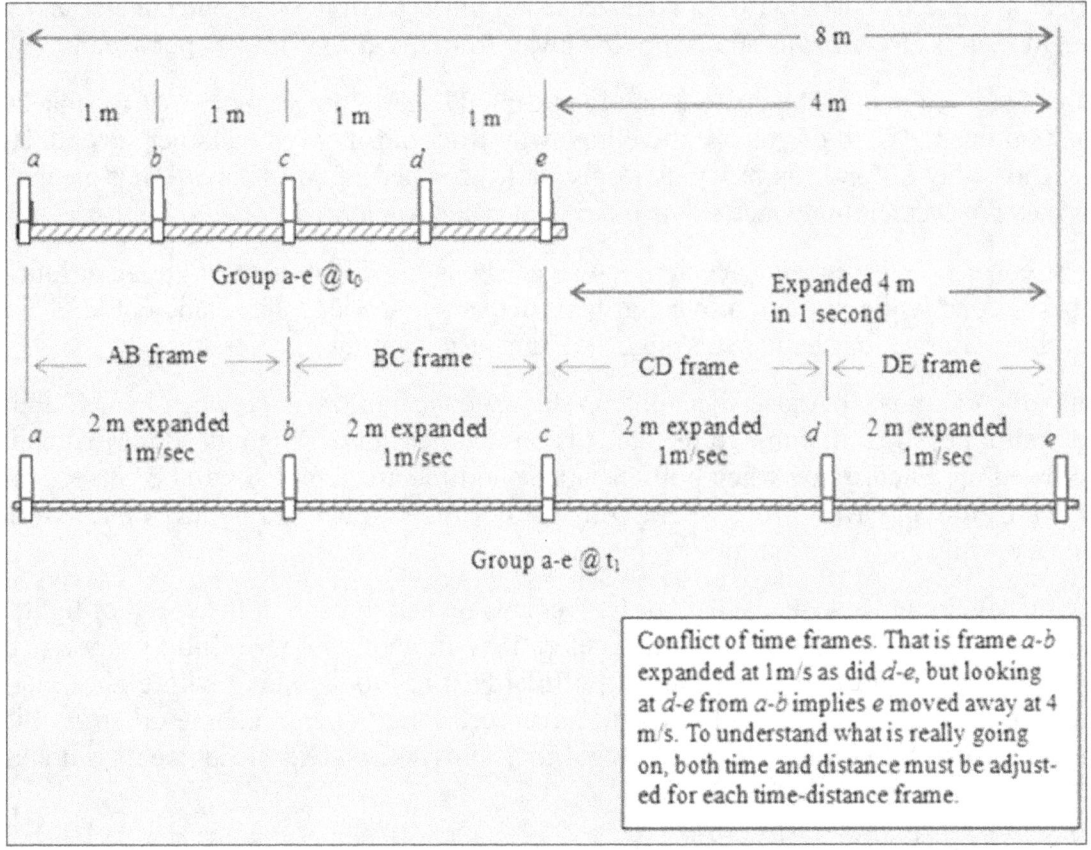

Referring to group *a-e* above, all clothes pins are equally spaced at one meter apart at time zero. Using a tool of your choice, remove the hook at e and stretch the cord four meters in one second. Secure it to a fixed post.

Let the mind game begin. The distance between *a* and *b* has doubled in one second as did *b-c* in time frame BC. Actually, the same is true for all time frames. An ant on pin *a* noticed that her friend on pin *b* had suddenly gotten smaller. She said politely, "Why did the distance between us double, and in only one moment? You have moved away from me at a speed of one meter per second."

But, an ant sitting on pin *c* noticed his girlfriend on pin *a* has moved from two meters away to four meters away in the same second and said so aloud. "No siree, honey," said the guy ant on pin *c*. "You've moved from two meters away to four meters away from me. Therefore, your speed was two meters per second."

"Hey, wait a minute, you idiots. The both of you," said the ant on pin *e* in ant speak. "One moment ago the distance between Anastasiya on pin *a* and me was four meters, and now it's eight meters. That means she is moving away from me at four meters per second."

"Both of you are nuts." The guy ant on pin *c* wanted back in the argument. "I'm smack dab in the middle, and both of you are traveling away from me at two meters per second. However, Anna May on pin *b* is moving at only half the speed of Anastasiya. That means that Anastasiya is accelerating since she is moving at twice Anna May's speed. So there."

Who is correct? Is Anastasiya accelerating since she is moving away from pin *e* at four meters per second while pin *c* is moving at two meters per second? The same could be asked about the quasars at our Universe's edge. Are they accelerating?

It's a game of sums. If Anastasiya adds up the distance and times between herself and her buddy on pin e, she will come to the conclusion that her universe is indeed normal and not so screwed up. Each frame when both distance and time are summed will be correct. However, who's moving away from whom will still be argued. Each ant believes their world is stationary while all others are moving away.

Since this alternate view states that our Universe is revealing itself at the speed of light, it is also expanding. But, is the expansion accelerating? In another 13.9 billion years our Universe will contain eight times the amount of matter it has today along with eight times the amount of space. Mankind and Earth will have been long gone, but some other intelligent being will come to the same conclusion that their universe is 27.8 billion years old and revealing itself at the speed of light.

However, there is a limit to light gathering equipment and to radio frequency gathering equipment. At what point will new intelligence become unaware that anything exists outside their electromagnetic gathering range? At what point will the Cosmic Radiation Background be too far away to detect? What will they make of their universe without those markers?

Some can argue that intelligent beings will be a few billion years smarter so they will figure it out. That probably will not happen. Our Universe has put a limit on how long any form of life can remain in existence. It is continuing to build itself at the outside edges while the inside grows old. Older generations of planets will continue to be blown to smithereens as solar systems are born and perish in various forms of disintegration.

Space cannot be created just to rationalize a theory, to justify matter exceeding the speed of light, or to explain how objects outside our vision exceed that restricting speed. The superforce alone is responsible for that constant, and the superforce alone is in charge of our Universe and its actions.

Furthermore, there is 84% more space than necessary to account for all known matter. Unless those failed subparticles are considered to be what created that extra space, our Universe will continue on doing whatever it is doing. But when the universe's pressurization ceases, and all that matter which created that extra space is brought into play, it may apply a braking force and bring the expansion to a halt. After that, a slow crunch may begin.

Don't worry about it. Either way, that amount of time for one or the other to occur is incomprehensible to mankind.

Note: An amazing thing happened a few months before the printing of this book. To this author it was almost as amazing as discovering life on another planet. An asteroid from another world entered and exited our solar system. The fact that exo-objects exists is not amazing, the fact that "We" witnessed it is amazing. Its name is Oumuamua (Hawaiian for "messenger" or "scout").

Further, this author believes that, somewhere within our solar system, rocks exist that were left over from planets that were circling the star that went supernova. It will be a great day when some future astronomers discover a rock *out there* that is over six billion years old. Even Oumuamua could be much greater than 4.3 billion years old. Just imagine. . . .

Chapter 20

Ice Ages

While many things contribute to global warming, this alternate view believes that precession of the equinoxes is the primary driver of climate change. Its period is roughly 25,920 years give or take. During the winter solstice, the earth is approaching its orbital perihelion, its closest approach to the sun. It is 147,100,000 kilometers from the sun and traveling at its maximum orbital speed at its closest point. Just past the summer solstice, the earth is farthest from the sun on or about July 4. Its distance is 152,100,000 kilometers and moving at its slowest pace.

Wiki 1 represents a current model of Earth's orbit.

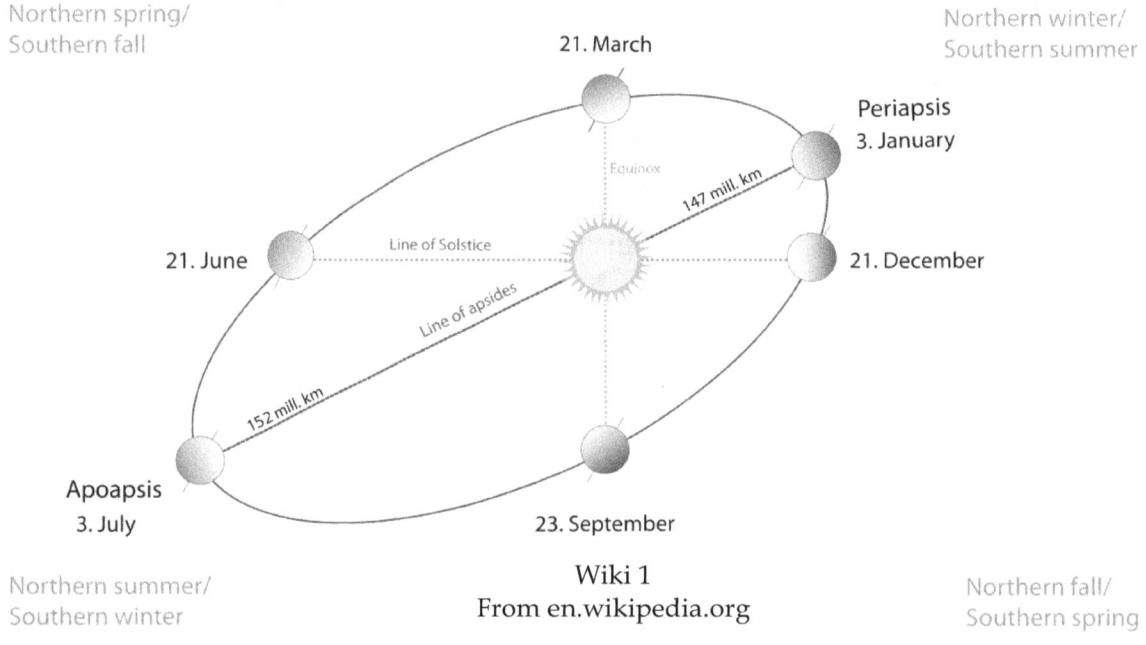

Wiki 1
From en.wikipedia.org

On average, the earth receives around 1368 watts per square meter at the top of the atmosphere. During the northern summer and near perihelion, ole Sol donates only 1323 watts per square meter. During the winter, the earth receives 1415 watts per square meter. Yep, the Northern Hemisphere receives more energy from the sun during its winter than its summer. That's a whopping 92 watts/meter^2 or nearly 7% more power than six months earlier. Is there any wonder that glaciers and the north polar ice cap melt? It seems rather natural that this should happen. But the atmosphere absorbs around 368 watts or 27% of the power, so only around 1000 watts hit the earth's surface, a little more in the winter; a little less in the summer. However, when it's winter in the Northern Hemisphere it is comfortable, right?

By the way, watt is the term for power. It is the ability to perform work, and it tells us how much work can be done with that amount of energy arriving from the sun. That is, we can turn on artificial lighting, operate an electric motor, heat leftovers, listen to a radio, or watch television.

Almost eight-hundred years ago, 1246 AD, the winter solstice aligned with the line of apsides (a line connecting the closest point to farthest point from the sun). The December solstice is now winding its way toward the aphelion. The meltdown is slowly coming to a halt and will soon begin to freeze again. There is a lag between the alignment and the time ice in the Northern Hemisphere begins to form into sheets. It's akin to the lag in the coldest part of winter taking place a few weeks after its solstice.

Move time ahead 12,188 years to exchange winter with summer and far enough along the ecliptic to superimpose the line of solstices over the line of apsides again as in Ice 1.

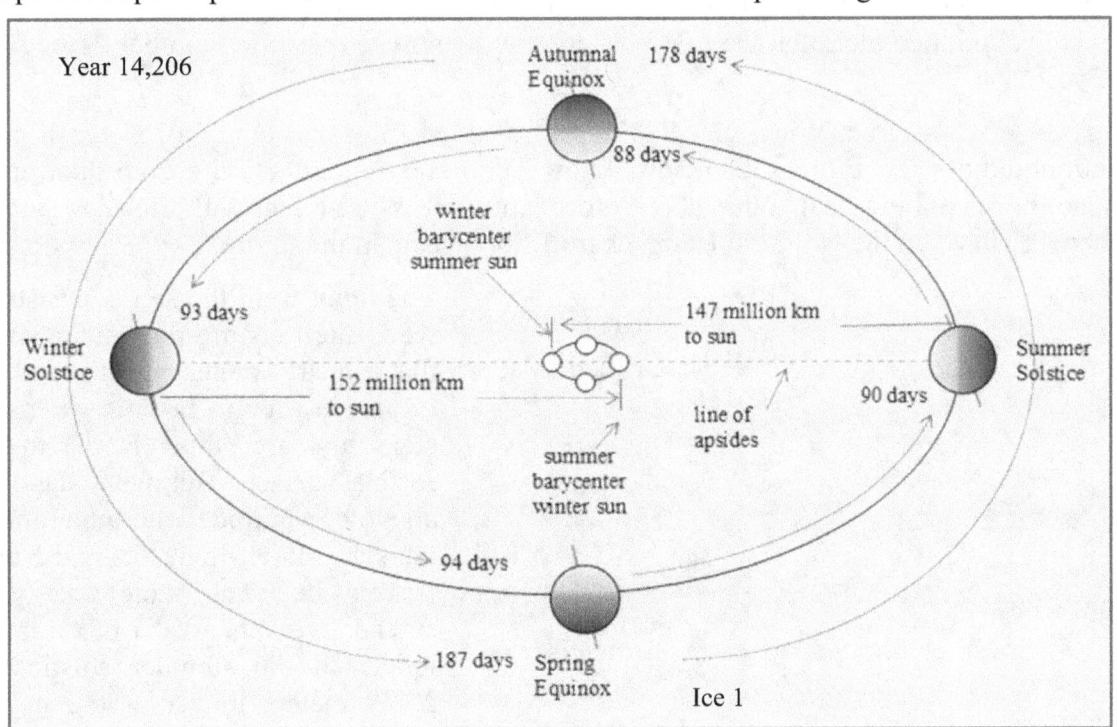

After time advances, the summer solstice occurs when Earth is closest to the sun, and the winter solstice is at Earth's greatest distance. Are winters comfortable at that point in future history?

Notice the variations of travel times between Earth's seasons: the winter solstice, the spring equinox, the summer solstice, the autumnal equinox, and back to the winter equinox. The total is 365 days (rounded), but the distribution is now greater on the winter side of the equinoxes. The moment the earth passes perihelion it begins to slow, and by the time it reaches aphelion, it is moving at its slowest pace; then it begins to gain speed again. The time spent on the winter side of the equinoxes is 187 days vs the summer side of 178 days. That's a week and two days longer and farther from the sun.

The surface energy received during winter in the tropics has changed from 1034 to 967 w/m^2 for a loss of 67 watts. That loss is equivalent to not being able to turn on a sixty watt incandescent light bulb. However, it still doesn't seem like very much, but it is also equal to 229 BTU's/hour. Power is energy used over a time period, so the energy must be applied over that same period, an hour in this case. What can a BTU do for you? What can that loss or gain affect with respect to ice over a period of one hour?

When added to the environment, it can raise the temperature of one pound of ice from -32° F to 32° F, and then melt that ice to 32° water and then raise the water another 21 degrees to 53° F.

The reverse is also true. When 229 BTU's are removed from that pound of water, it goes from a liquid of +53° F to a -32° F solid. However, removing 229 BTU's from the tropics may not be such a big deal. What about other latitudes? What effect will a loss or gain of heat energy have on the Northern Hemisphere 12,000 years in the future?

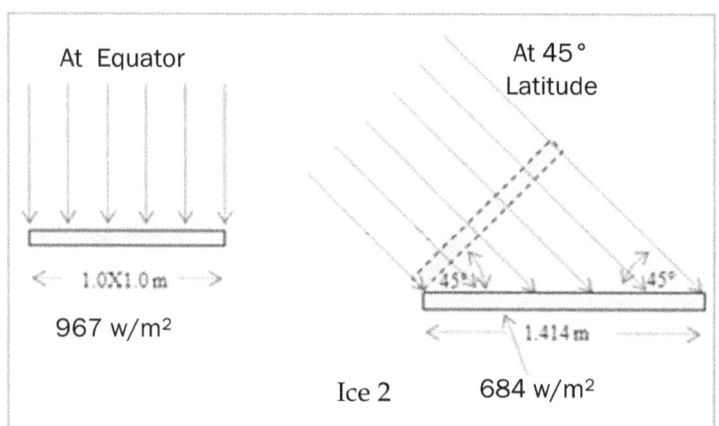

The input from the sun is measured when the rays are perpendicular to that square meter. That's in the tropics at a point in time when the sun's rays are perpendicular to the earth's surface. But what happens above 45° latitude where the northern states like Minnesota and North Dakota lie? The same energy is spread over an area 1.078 meters longer at the summer solstice to 2.728 meters longer at the winter solstice. When the sun is over the equator for that 45°, the energy is spread along a strip one meter wide by 1.414 meters long or 1.414 m^2. See Ice 2. The range is from 1.078 m^2 to 2.728 m^2 over a one year period. And as the area changes, the power varies from 900 w/m^2

during the summer to 354 w/m² during the winter. A one horsepower electric motor requires 740 watts to operate, so if the atmosphere is dry in the summertime that's enough power to drive a large swamp cooler and have some left over for watching television. But there is something missing from the above rationale—atmospheric attenuation.

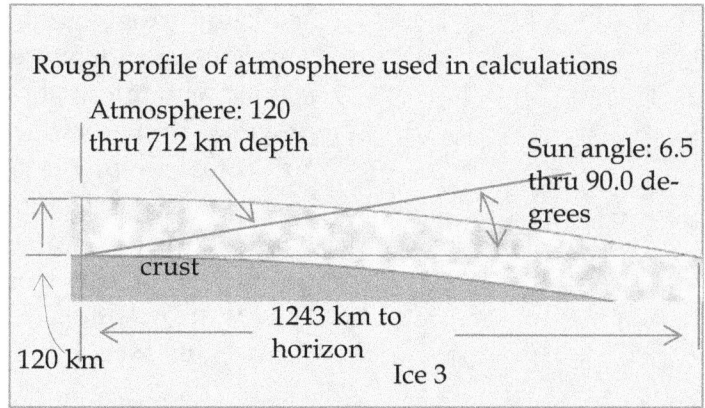

Ice 3

Not all that energy makes it to the ground. It must travel through a thicker layer of air brought on by the sun's ever changing altitude in the sky. The thicker the air, the more molecules the energy must penetrate. The extra air reduces the energy before it ever makes its way to the ground. See Ice 3.

When both atmosphere and sun elevation losses are accounted for, the new range begins at 834 w/m² when the sun is at the Tropic of Cancer, so forget about the TV. That 834 quickly drops to 153 w/m² when the sun is over the tropic of Capricorn. In addition, as the sun dives to the south that 153 watts is applied to the earth's surface for a shorter time. That's cold enough to freeze 'em off the proverbial brass monkey. Ice will accumulate several hundred meters thick over a 13,000-year period.

For a comparison of temperature at the 45th parallel, when the sun is at the winter solstice, today's winter surface temperature is 33.0° F. In the year 14,166, it will be -3.2° F. The total gone missing is 36.2° F. That missing energy will cost a lot in wheat and other grain production, if there is any workable land in that area at all. Although the actual missing power is only ten watts, it will be devastating over a few-thousand-year period.

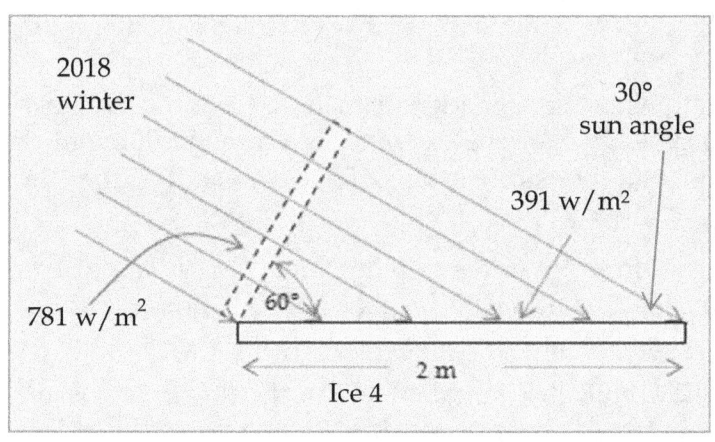

Ice 4

But humans have lived through it before. The Neanderthal did quite well until it became a little too warm for them to adapt. They will not come back, but will someone else take their place?

It doesn't take a lot of imagination to realize that the sun provides under half the energy north of 60° latitude including atmospheric attenuation, and it's available for a much shorter time. When the sun is overhead at the Tropic of Capricorn the atmosphere is 120 km thick at that location. At a winter sun elevation of 21.5°, the atmosphere is close to 306 km thick, and at 6.5°, it is around 712 km in depth. That's a lot of reduction due to the thicker air alone.

Models of Future Seasons

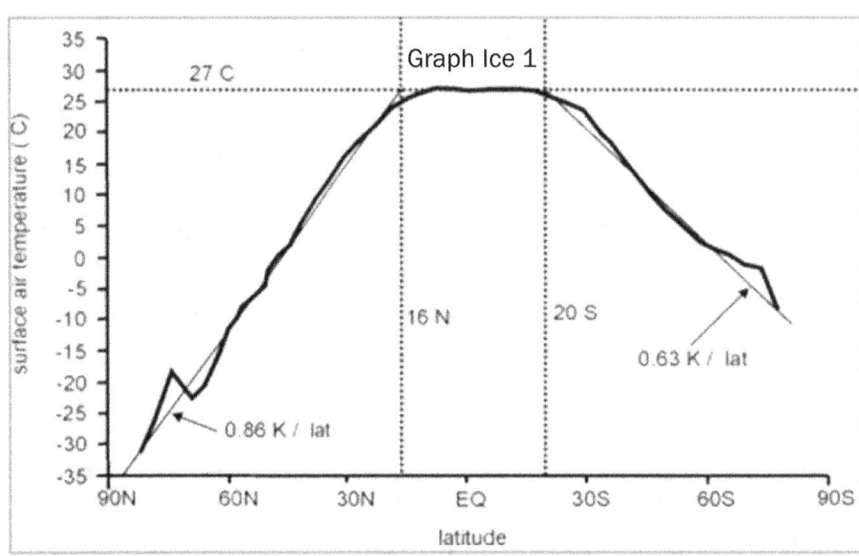

Modeling for current and future temperatures is done with information provided by the surface air temperature chart presented by Graph Ice 1. Thanks to Professor Geerts of the University of Wyoming for allowing its use.

There is a great difference between how the same energy affects the Northern and Southern Hemispheres, water being the driver of those differences. The slope of the north latitudes changes from being sturdy at the equator to having a 0.31° Kelvin (absolute zero) per degree latitude at 07N. It continues at that rate until it reaches 16N at which time it steepens to 0.86° K per degree latitude. The southern latitudes hang on to rather level grade from the equator until 20S. It then bends slowly downward until it settles in at 0.63° K per degree latitude until it touches Antarctica. More on that shortly.

Several models are chosen, and information will be presented in both tabular and graphic form.

A major component used in the calculations is atmospheric attenuation. Ice 3, on previous page, represents the depth of atmosphere used. Sun angles from zero through 90 degrees are calculated. Both atmospheric attenuation and sun elevation for each parallel from the equator to both poles are considered.

The illustration displays the sun elevation from 90 down to 6.5 degrees, but it can just as easily set on the horizon. Atmosphere depth is 1243 km at that point. It doesn't matter whether it's a setting sun, rising sun, or southern sun. The attenuation is the same.

Surface temperature tells how hot or how cold that surface feels to the touch. It doesn't matter its makeup: water, ice, snow, rocks, or any other noun. It is not the same as ambient air temperature. However, it does affect everything around it.

Table Ice 1 displays current temperatures and power delivered by the sun at the latitudes indicated. Both tropics are rounded down to 23° in tabular and graphic form.

Table Ice 1									
Northern Hemisphere Summer					Southern Hemisphere Winter				
Lat	C	F	Power	Sun°	Lat	C	F	Power	Sun°
0 N	8.2	46.8	822	N@67	0 S	8.2	46.8	822	N@67
15 N	10.9	51.7	950	N@82	15 S	8.2	46.8	606	N@52
23 N	17.2	63.0	967	90	23 S	6.5	43.7	471	N@44
30 N	11.2	52.2	952	S@83	30 S	2.1	35.7	356	N@37
45 N	-1.7	29.0	834	S@68	45 S	-7.4	18.7	143	N@22
60 N	-14.6	5.8	622	S@53	60 S	-16.8	1.7	21	N@07
75 N	-27.5	-17.5	372	S@38	75 S	-29.6	-21.2	0	-8
90 N	-40.4	-40.7	155	S@23	90 S	-56.0	-68.7	0	-23

Practice reading the table under Northern Hemisphere Summer. For folks on the equator, the sun is north at 67°, for folks on the 15th parallel, the sun is north at 82°, and for folks under the Tropic of Cancer, the sun is directly overhead.

The Southern Hemisphere Winter is read as before. Folks on the equator see the sun north at 67° elevation, folks at 15 S see the sun north at 52° elevation, and the folks under the Tropic of Capricorn see the sun farther north at 44° elevation.

Winter/Summer statistics for every 15 degree north/south latitude are displayed in each row, the odd-balls being the Tropic of Cancer and the Tropic of Capricorn. They are available to glean a bit more information. The minus prefix for southern latitudes is absent to avoid confusion with negative sun elevations where the minus sign indicates the angle below the horizon. In Table Ice 1 at 75 S the sun is beneath the horizon -8 degrees, at 90 S it's under by -23 degrees.

Currently, the Northern Hemisphere summer and the Southern Hemisphere winter take place when the earth is farthest from the sun. That's why the winters at the South Pole are so cold. In 13,000 years, they will have already switched.

Universe's Mathematics

The curve below was generated on the fly as tabular data was calculated for the northern summer and southern winter for table Ice 1. The graph was generated for every degree north and south from the equator to both poles. The dots along the x axis designate latitude for every 15 degrees with the corresponding temperature in C along the y axis. Exceptions are the Tropic of Cancer and the Tropic of Capricorn. They are rounded from 23.4 to 23° even. Their significance is to mark the location where temperature peaks during the summer months.

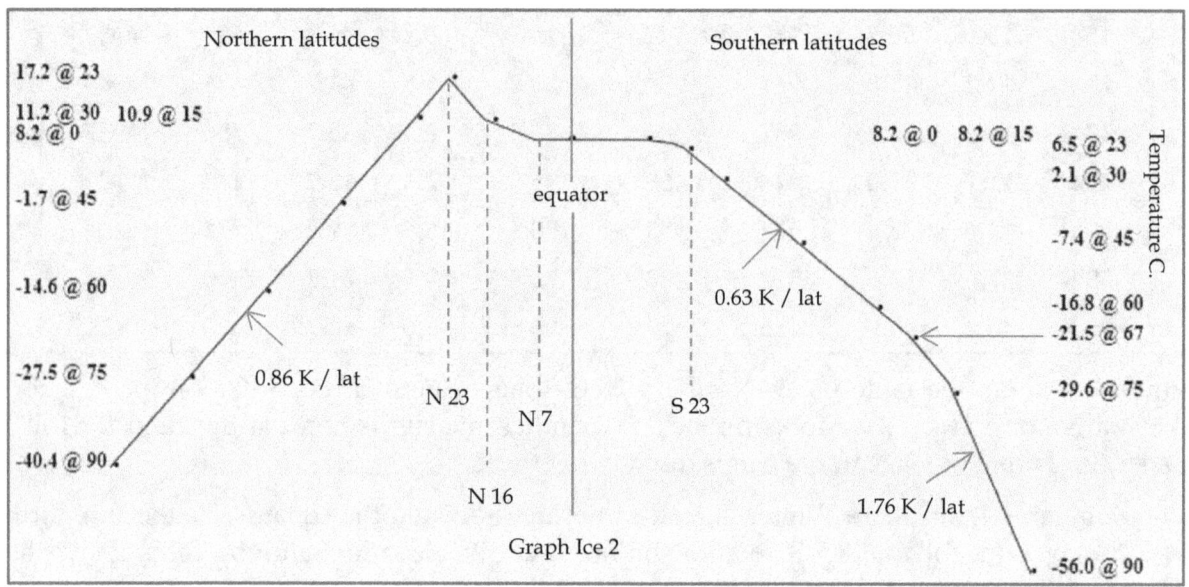

Graph Ice 2

For example, the numbers 17.2 @ 23 in the upper left corner mean 17.2° C. @ N 23. Another exception is the Antarctic Circle. It is identified by -21.5 @ 67 which means the temperature is -21.5°C. at 67 S. Not all rates are shown on the graph, so knowing all that, let's plot the graph in text to include all the numbers.

Beginning at the equator at high noon and going north, the temperature and latitudes are plotted for the moment the sun is at its highest point of the day. The continuous line represents every degree N and S latitude, and the dots mark every 15° NS latitude with the aforementioned exceptions.

The plot begins at the equator with a temperature of 8.2°C. (third row down from upper left corner), and as it makes its way north it begins to rise at a rate of 0.31° K / lat around the 7th parallel. It continues at that rate until the 16th parallel. At that point, the incline jumps to 0.86 and remains there until the sun is directly overhead at N 23 where it reaches its peak of 17.2°. There, the incline turns downhill, and the rate of decent remains at -0.86° K / lat.

Going south, the plot begins at the equator with the same temp of 8.2. It remains level until after 15 S where it bends slightly downwards. At 20 S, it continues its downward motion, and at 23 S, it settles in at a rate of -0.63° K / lat. The rate remains at that level until another exception is reached at the Antarctic Circle where it steepens to about -0.91° K / lat. At that position, the eastern half of Antarctica lies beneath 67 S, but a lot of water at the western end of the continent won't allow the fall to get any steeper. Finally, at the 75th parallel it dives at -1.76° K / lat.

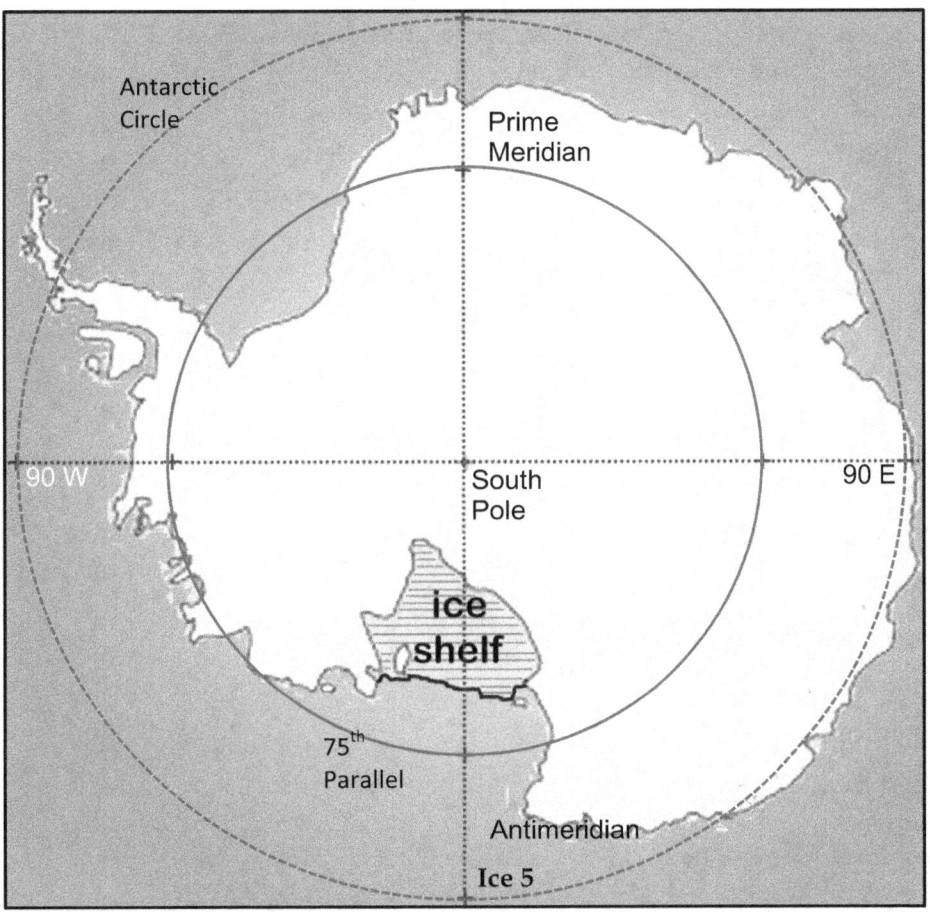

The overall rate of temperature change for the Northern Hemisphere is steeper than the southern half of the earth. Water, that's the reason the southern rate is flatter than the northern. As the plot moves southwardly, just under 2/3rds of Earth is covered with water until approaching Antarctica. Again, land instigates the slope angle as Ice 5 implies.

On the first day of southern summer with the plot at the Antarctic Circle, about 66% of the area under the sun is land which just about matches the Northern Hemisphere's land to water ratio. At 75 S, the ratio exceeds the northern portion, and there's nothing to prevent temperatures from taking a steep incline. When the sun moves back over the Tropic of Cancer, 98% of Antarctica is in the dark. The little finger sticking out at the western edge, and shelves just outside the 67th parallel on the eastern edge of Antarctica, are the only locations to receive the sun on that first day of southern winter. And it is on the horizon. Looking at Ice 5, percentages are just another WAG.

The South Pole is on land surrounded by water; the North Pole is on water surrounded by land. That makes a huge difference in temperature gradients. Wiki 2 modified from Wiki Commons is provided for a quick contrast.

The dashed circle represents the Arctic Circle.

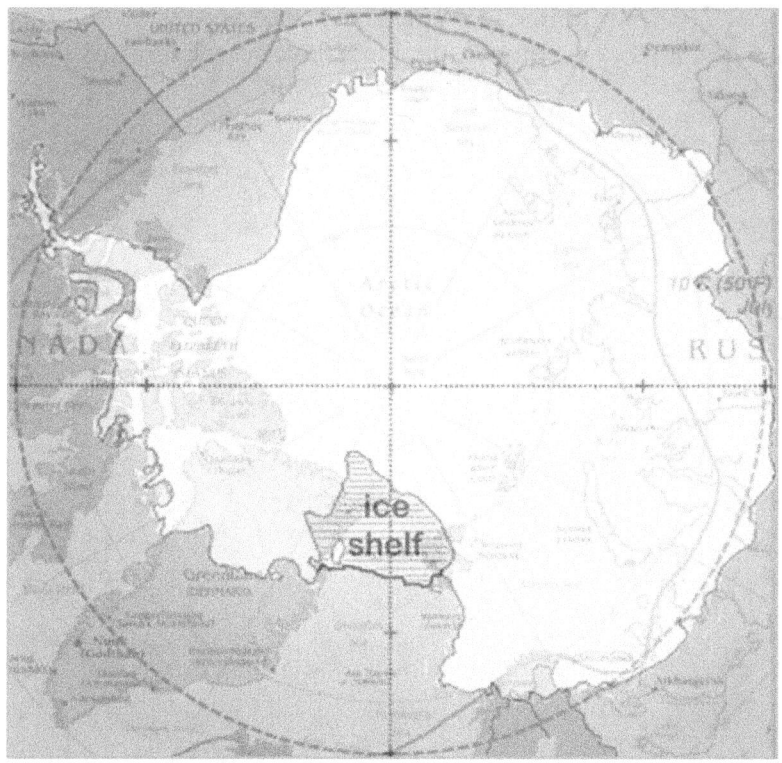

A couple more pictures before you go. In the image on the left, Antarctica has been superimposed over the North Pole for a quick comparison of land to water ratio of South Pole vs North Pole.

If sea ice were included in the image, everything under the Arctic Circle would be either solid ice or land. Nevertheless, the continent still covers a lot of north Russia, Greenland, and Canada's northern islands.

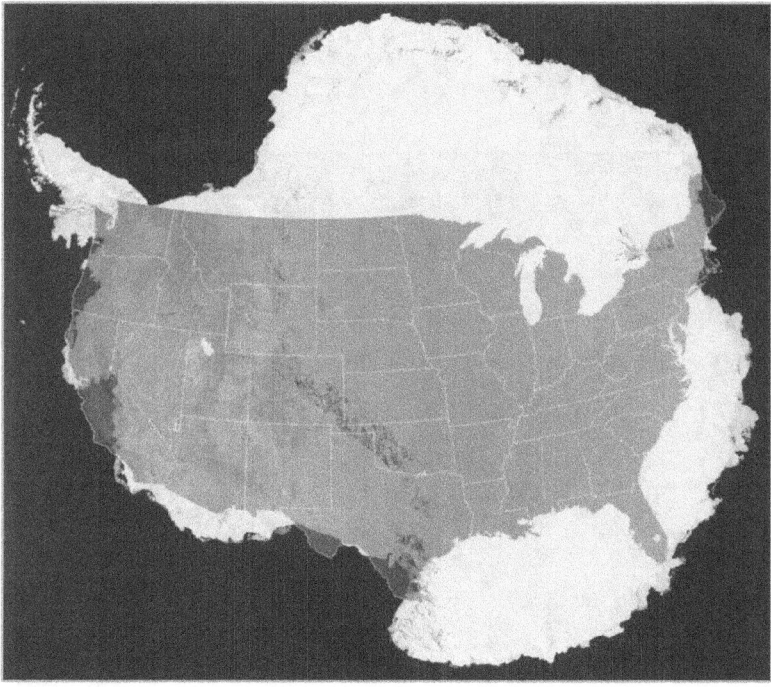

The image on the left is from NASA, and it shows the United States superimposed on Antarctica. The 5th largest continent extends up into Canada about a thousand miles and down into the Gulf of Mexico. Mile-thick ice covers most of Antarctica. It is probably close to being what it was like over 13,000 years ago in North America.

We can look forward to those huge ice sheets covering the Northern Hemisphere again, maybe not as high, but they will be thick.

Curve Ice 2 contains slopes for year 2018 N Summer S Winter and year 15,018 N Summer S Winter while Table Ice 2 gives a quick tabular comparison between temperatures of those summers.

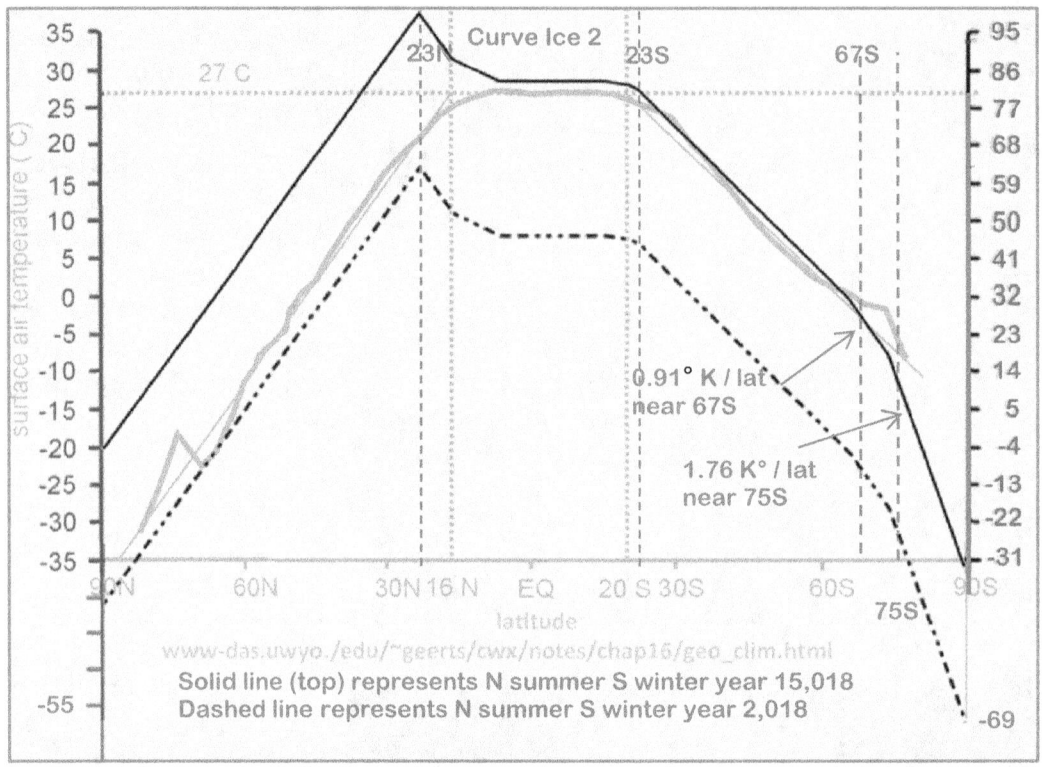

Solid line (top) represents N summer S winter year 15,018
Dashed line represents N summer S winter year 2,018

Table Ice 2									
North Hemisphere Summer 2,018					North Hemisphere Summer 15,018				
Lat	C	F	Pwr	Sun°	Lat	C	F	Pwr	Sun°
0	8.2	46.8	822	N@67	0	28.5	83.3	870	N@67
15	10.9	51.7	950	N@82	15	31.2	88.2	1006	N@82
23	17.2	63.0	967	90	23	37.5	99.5	1024	90
30	11.2	52.2	952	S@83	30	31.5	88.7	1008	S@83
45	-1.7	29.0	834	S@68	45	18.6	65.5	883	S@68
60	-14.6	5.8	622	S@53	60	5.7	42.3	659	S@53
75	-27.5	-17.5	372	S@38	75	-7.2	19.0	394	S@38
90	-40.4	-40.7	155	S@23	90	-20.1	-4.2	164	S@23

The original surface air temp has been placed in the background as a basis. The Fahrenheit pole was added for familiarity along with the rate changes near the South Pole.

Although this topic is about Ice Ages, there is also a summer consequence to deal with. The heat reaches its peak of 63.0° F at the Tropic of Cancer (23N) of for our current summer season, but after seasons are switched 12,148 years from now the peak will hit 99.5° F. Yet ice will build in the Northern Hemisphere because the earth swoops by the sun swiftly during that summertime giving too little time to melt the winter's accumulated ice.

The next two representations will indicate why.

Curve Ice 3 illustrates the current peak temperature in the Southern Hemisphere during its

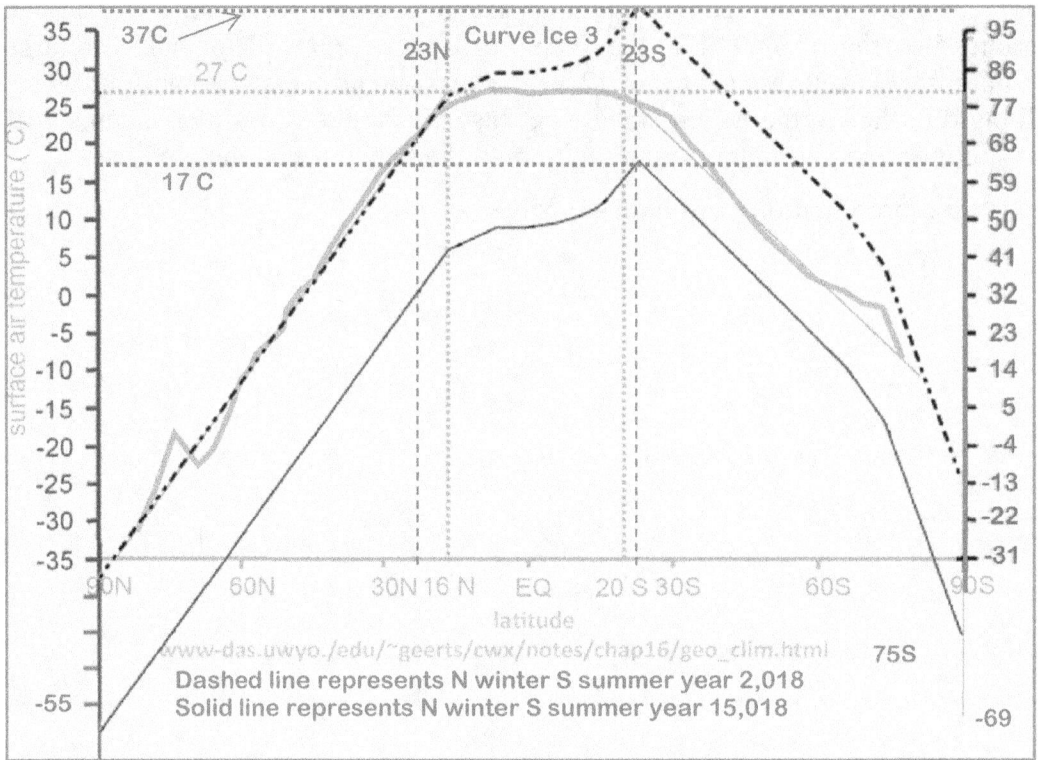

summer and the north's winter. Compare -38.2 vs -58.2 C at the North Pole in Table Ice 3. For Fahrenheit it is -36.7 vs -72.8.

Table Ice 3									
North Hemisphere Winter 2,018					North Hemisphere Winter 15,018			
Lat	C	F	Pwr	Sun°	Lat	C	F	Pwr	Sun°
0	28.5	83.3	879	S@67	0	8.4	47.1	822	S@67
15	25.8	78.4	648	S@52	15	5.7	42.3	606	S@52
23	19.5	67.0	504	S@44	23	-0.6	30.9	471	S@44
30	13.4	56.2	381	S@37	30	-6.6	20.0	356	S@37
45	0.5	33.0	153	S@22	45	-19.5	-3.2	143	S@22
60	-12.4	9.8	22	S@07	60	-32.4	-26.4	21	S@07
75	-25.3	-13.5	0.0	-8	75	-45.3	-49.6	0	-8
90	-38.2	-36.7	0.0	-23	90	-58.2	-72.8	0	-23

When far future temperatures are stacked next to present temps, they show symmetry between opposite seasons. The solid lines represent north winters while dashed lines represent north summers. The top solid line is for year 2,018, bottom is for 15,018. Bottom 15,018 is symmetrical with dashed N Summer S Winter 2,018. Top solid is symmetrical with dashed N Summer S Winter 15,018.

The seasons have clearly flipped, yet some people declare there is no change in climate.

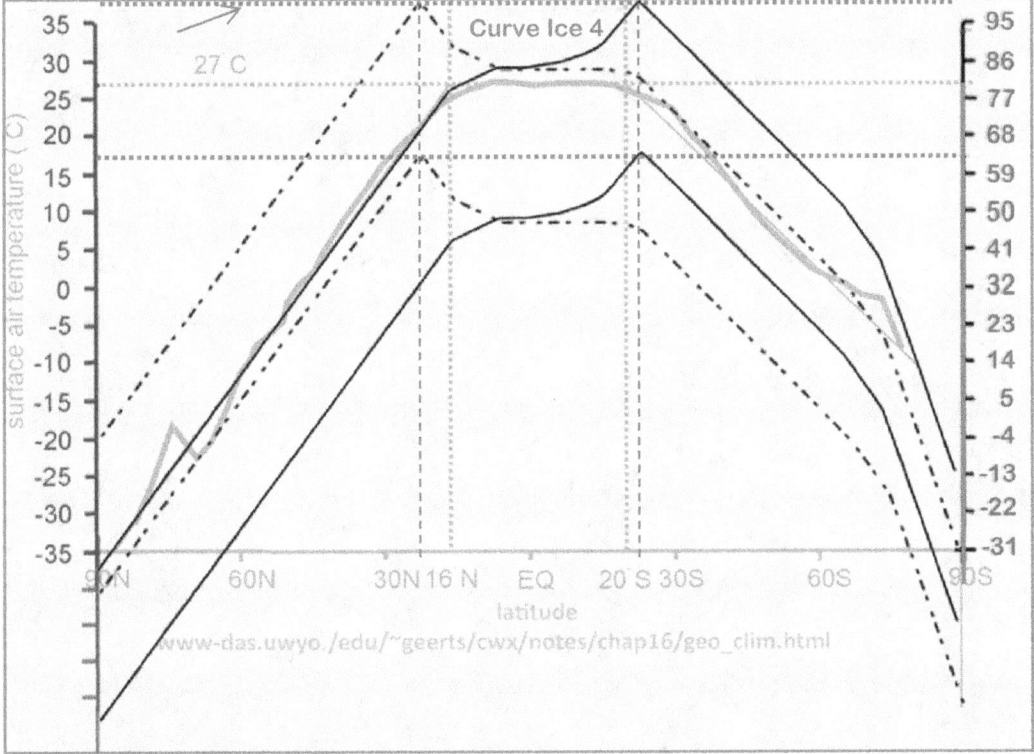

There is, and it is due to water vs land.

Put yourself in the year 15,018 when ice covers the Northern Hemisphere from about 40° latitude to the North Pole. That includes Europe and Asia. Instead of worrying about going on vacation or what's for entertainment tonight, those people will be more concerned about production of nourishment because the Southern Hemisphere contains less food productive land than the Northern. Unless the Sahara Desert changes to accommodate wheat and other grains, the whole world will be on a strict diet.

Universe's Mathematics

Earth's orbital eccentricity (*e*) goes awry every hundred-thousand years or so. An ellipse with an eccentricity of zero is a circle. Earth's minimum *e* of 0.005 puts it close to a circle. Currently its *e* is 0.017 with a maximum of 0.058, the latter being a long ellipse with a shorter distance of 140,900,000 km to the sun, and a longer distance of 158,300,000 km at apsis. A difference of 17.4 million km makes for extremely hot summers and extremely cold winters in the Northern Hemisphere. Jupiter plays a large role in squashing the orbit, but other things affect it as well. These changes are all part of the Milankovitch cycle.

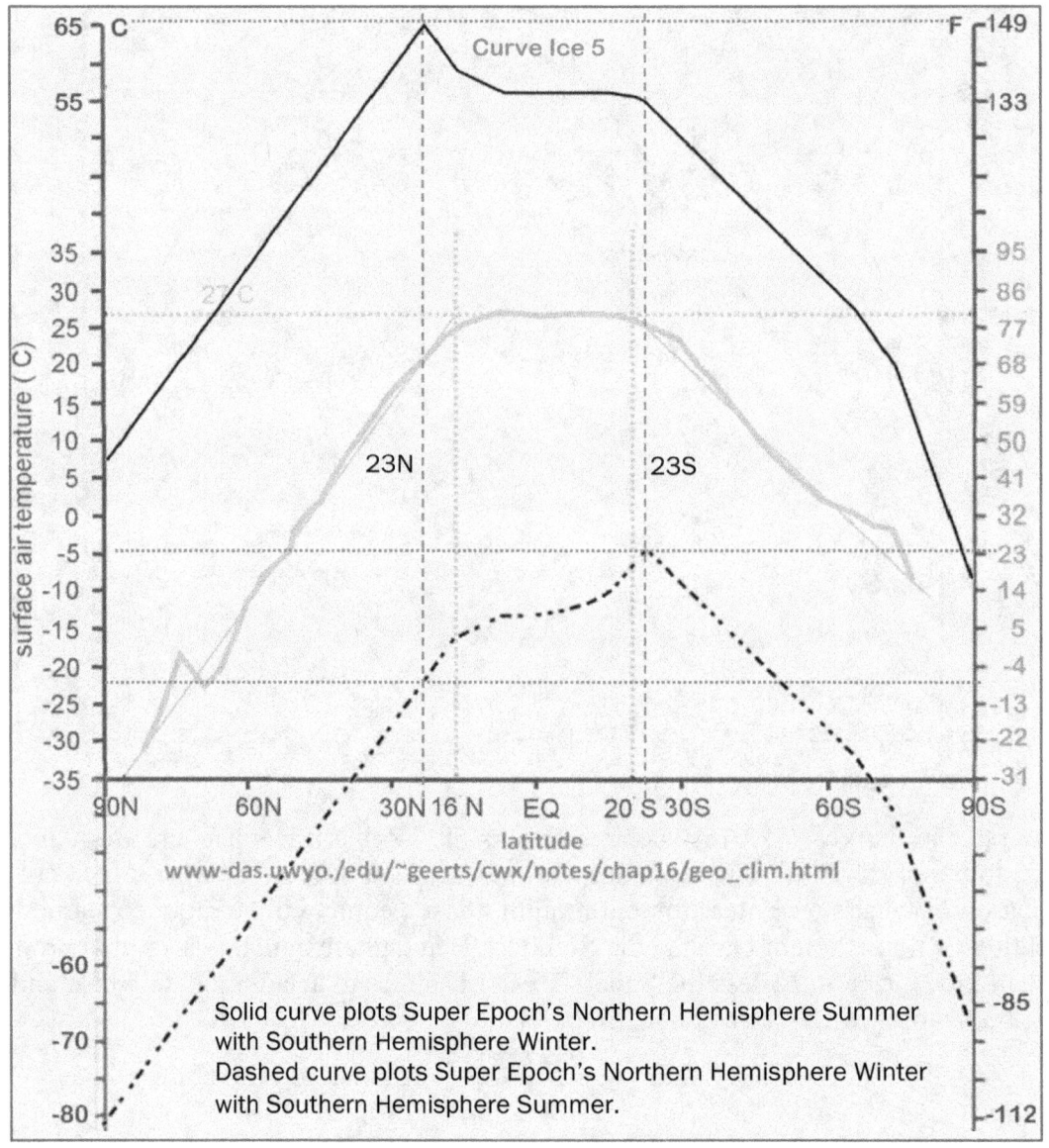

The window into that super epoch had to be modified greatly in height to accommodate the extreme heat and cold temperatures for a comparison to today's surface temps.

Data for the hot season of that Super Epoch is in Table Ice 4 while the plot is the solid line in Curve Ice 5. The name Super Epoch seems to apply because everything about that time period is just that, super. For sure, mankind will definitely be miserable during that few thousand years if it survives at all. That is, as we know mankind today.

Table Ice 4 SuperEpoch									
North Hemisphere Summer					Southern Hemisphere Winter				
Lat	C	F	Pwr	Sun°	Lat	C	F	Pwr	Sun°
0	56.2	133.2	957.6	N@67	0	56.2	133.2	957.6	N@67
15	58.9	138.0	1106.8	N@82	15	56.2	133.2	705.7	N@52
23	65.2	149.4	1127.0	90	23	54.5	130.0	549.4	N@44
30	59.2	138.6	1109.4	S@83	30	50.1	122.1	415.3	N@37
45	46.3	115.4	972.1	S@68	45	40.6	105.1	166.7	N@22
60	33.4	92.1	724.9	S@53	60	31.2	88.1	23.9	N@07
75	20.5	68.9	433.7	S@38	75	18.4	65.2	0.0	-8
90	7.6	45.7	180.4	S@23	90	-8.0	17.6	0.0	-23
Table Ice 5 SuperEpoch									
North Hemisphere Winter					Southern Hemisphere Summer				
Lat	C	F	Pwr	Sun°	Lat	C	F	Pwr	Sun°
0	-13.8	7.2	758.8	S@67	0	-13.7	7.3	758.8	S@67
15	-16.5	2.3	559.3	S@52	15	-10.7	12.8	877.0	S@82
23	-22.8	-9.1	435.3	S@44	23	-5.0	22.9	893.0	90
30	-28.9	-19.9	329.0	S@37	30	-9.4	15.0	879.0	N@83
45	-41.8	-43.2	132.0	S@22	45	-18.9	-2.0	770.2	N@68
60	-54.7	-66.4	19.0	S@07	60	-28.3	-19.0	574.4	N@53
75	-67.6	-89.6	0.0	-8	75	-41.1	-41.9	343.6	N@38
90	-80.5	-112.8	0.0	-23	90	-67.5	-89.5	142.9	N@23

Table Ice 5 displays information pertaining to the dashed curve that plots the cold temperatures at latitudes zero thru 90N and zero thru 90S. Both curves are superimposed on the original basis curve with the aforementioned modifications. It had to grow tall to accommodate the punishing heat and cold temperatures.

Contrary Evidence

After all that talk about heat energy at north and south latitudes with the northern being colder, there is evidence that shows a contrarian view of the 2,018 seasons—the poles. Below is the average temperature at both poles with the South Pole much colder: one reason is that it is located 9,300 ft above sea level. Even after the altitude is accounted for, it is still around eleven degrees colder on the average. Another reason, it's farther away from the sun during the winter months than the North Pole.

Again, the real cause of the variance is water. The North Pole is on water; the South Pole on land, many square miles of it. As mentioned before, the Northern Hemisphere is mostly land with Russia and Asia containing the most by far. However, even the contrary view still attests that the greater distance during winters provide less heat energy than the close-up summers.

Be careful when finding the difference between temperatures of north and south. The conversion for altitude was performed at 3.5 degrees/1000 feet on the F scale, and then F was converted to C. If comparing directly by Centigrade, errors come quickly because that is a different formula for altitude correction.

Temperatures rounded

Time of year	Average (mean) temperature		Altitude correction for South Pole	Difference
	North Pole elevation 0 ft	South Pole elevation 9301 ft		
Summer	32° F (0° C)	−18° F (−28.2° C)	14° F (−10° C)	18° F (-8° C)
Winter	−40° F (−40° C)	−76° F (−60° C)	−44° F (−42° C)	-4° F (-20° C)
Average	-4° F (-20° C)	−47° F (−44° C)	-15° F (-26 C°)	-11° F (-24° C)

Another important part of Ice Ages is a wandering barycenter. We're forgoing any reference to temperatures because of data indigestion.

Wandering Barycenter

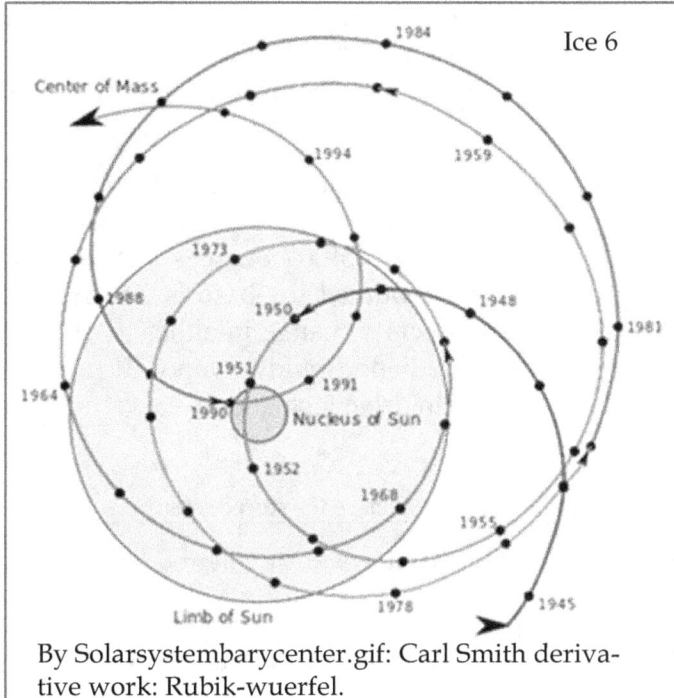

Ice 6

By Solarsystembarycenter.gif: Carl Smith derivative work: Rubik-wuerfel.

Our Universe sets up other conditions that change Ice Age cycles: some lasting millions of years while others last several thousand years, even hundreds. See Milankovitch cycles on the web for that information.

For example, say an extreme cold state occurs around 15,018 AD which means ice has already been accumulating for over 4,000 years. That ice will begin thawing somewhere near 19,000 AD and continue for another 2,000 years. After that, it will melt at an outlandish rate as the earth's winter solstice gets nearer the sun. After the solstice passes periapsis, the melt rate will begin to slow.

Before going further, a review of barycenters is necessary. During the subject of orbits, we learned about the barycenter when only two objects were involved. Remember the seesaw. Each time another object is thrown into the mix, the center changes. The complexity with each new planet, major or minor, grows rapidly. Hats off to the mathematicians involved in those calculations: then there is the asteroid belt, the Kuiper belt, the Oort cloud, wandering stars, and Planet 9, that recently discovered giant that revolves around the sun every 20,000 years or more. When everything affecting the barycenter comes into play, it is no longer on a plane. It acts more like it's inside a sphere that has been flattened with the solar system plane in charge. Well, all the planets taken together aren't really on a plane. Maybe each paired with the sun is, and then they can be combined into a very thick plane.

Imagine that it happens during the time implied by Ice 1 (p132) in the year 14,206. When an adjustment is made to include that disruption to the barycenter, Earth is 153,400,000 km from the sun with only 950 watts/m^2 worth of heating ability during winter. And that falls on the surface at the equator. That's a direct loss of 8.2% power from the sun. The winter temperature at latitude 45 will be 3.2° F, a far cry from today's winter temperature of 33.0 F.

Further understanding of how power and energy affect temperature is required, so a reference to energy stored in water is below. It begins at a temperature where no energy exists at all; that temperature is zero degrees Rankin. That is a minus 459.67 degrees F. When the energy to raise ice from 0° R (-459.67° F) to 32° F is considered the base is 246 BTU's. Any removal or addition to the energy stored in water affects a change in either its state or its temperature. Examples of state changes are ice to liquid and liquid to vapor. It requires much more energy to affect a change in state than it does to drive a change in temperature.

Table Ice 6
Energy to raise water temp from absolute zero (R) to 32° F is base for stored energy

energy stored in 1 lb. of solid water	@ 32° F 246 BTU's	0.5 BTU raises 1 lb. Ice 1 degree F
latent heat of fusion to melt ice and maintain 32° F	144 BTU's	changes solid to liquid; temp remains at 32°
energy stored in 1 lb. of liquid water	@ 32° F 390 BTU's	1 BTU raises 1 lb. water 1 degree F
energy stored in 1 lb. of water	@ 212° F 570 BTU's	
latent heat of vaporization	970 BTU's	changes liquid to vapor; temp remains at 212° at standard atmosphere
energy stored in 1 lb. of water vapor (steam)	@212° F 1540 BTU's	can be super-heated under pressure
energy stored in atmospheric water vapor depends on history, i.e. what drove the change?	sublimation: 1360 BTU's evaporation: * from 1361 to 1540 BTU's	* depends on temp of water before evaporation

Let's put water and ice in perspective: a pound of water will fit inside a container slightly larger than a three inch cube—27.7 in^3, and a pound of ice is 28.8 in^3, so a pound of ice will fit snuggly inside a pint jar. Water will have a little more room to spare. On the metric scale, a liter is 61.02 in^3, so a lb. of water or ice is just under ½ liter.

Realizing that water vapor contains up to 1540 BTU's per pound makes it a little more meaningful when Dallas Raines, or any other meteorologist, speaks about a storm front coming down from Alaska or a hurricane building off the African coast. When they refer to the energy in the system, they are talking about billions of BTU's lying in wait . . . just lying there waiting for the opportunity to convert a vapor back to a liquid or a solid. With that comes a far reaching change in temperature of the atmosphere followed by an upheaval in turbulence, wind speed and direction, cyclonic movement of air, and megatons of water (liquid or solid or both) sweeping down on Earth's crust. It's really only the atmosphere undergoing a cleaning process—ridding itself of moist impurities, it is.

Non-contaminated water is an insulator. The newly formed droplets cannot conduct electrons, so static electricity runs amok until it becomes powerful enough to break down air molecules and make its way to the ground, another cloud, or even upward to an oppositely charged field of atmosphere. "Just get rid of these excess electrons," is the command.

There may even be another electron producing process. When tons of water vapor loses energy all at once, some electrons may gain enough energy to free themselves from the hydrogen/oxygen bond. After that, they may wander aimlessly in the atmosphere until the whole cloud becomes so negatively charged the field will overpower the air and tear it apart as those excess electrons make their way to neutral territory.

This is just a question. Why would anyone believe that after lightning travels two or three miles through air, another six inches separated from the ground by rubber would halt its progress? Okay, it just bothered this author as a kid, so it had to be asked.

Earth's Ice Ages are ongoing. They are either building or thawing. Some are much greater than others, and never are they everlasting. Our current Ice Age is in the meltdown process. It began thawing over 13,000 years ago and is still underway, but soon it will reverse itself and begin building again, perhaps another hundred years or more. The winter solstice passed the line of apsides in 1,246 AD, so the meltdown will soon halt and the freeze will begin again.

It is going to be very uncomfortable any place north of the equator around 12,000 years from now.

Readers who may be interested in learning more about Planet 9 please check out the following link.

Evidence for a Distant Giant Planet in the Solar System

Batygin, Konstantin and Brown, Michael E. (2016) *Evidence for a Distant Giant Planet in the Solar System.* Astronomical Journal, 151 (2). Art. No. 22. ISSN 0004-6256. http://resolver.caltech.edu/CaltechAUTHORS:20160120-093551312

Chapter 21

Shockwave

How the universe uses the power of two to achieve stability has been covered. Actually, it can be shown that any number can be a property of two dimensions or more. Something like, length times width results in an area which is a 2D result. Another example may be the number nine. Without context there is no way to know, but if has dimensions of 3 by 3 by 3 we know it represents a 3D object, a box of some sort. And there are times when it is important to deal with mixed dimensions: 2D * 3D.

For example, Boyle's law states that for a given volume of a gas, the product of its initial volume times it initial pressure is equal to its new volume times its new pressure when temperature remains the same. That is to say, $P1V1=P2V2$ where P1 is pressure at stage one, V1 is volume at stage one, and P2 is pressure at stage two, and V2 is volume at stage two. Then again, someone may ask, 'P1V1 what?' The what in this case, is pounds/in^2 * in^3, which ends up as pound-inches, or switched around as inch-pounds. Some mechanics will recognize this as similar to torque. It's just a product of two different measurements: in this case, PSI*volume where PSI is the 2D factor and volume is the 3D factor. To avoid confusion most engineers and scientists consider such numbers as scalers. This saves a lot of cancelling and squaring of numbers.

For instance, a cylinder having 102 in^3 when the piston is fully extended with a swept volume of 90 in^3 will have a volume of 12 in^3 when its piston is at the top with a compression ratio of 8 ½ to 1. Given these specs, what is the absolute pressure at sea level after one cycle from top to bottom back to top? This is also known as intake-compression stroke. The function of an internal combustion engine is not the topic, just the first cycle. Hopefully it will lead to a better understanding of the development of a sonic shockwave.

Rounding sea level to 15 PSI (absolute), when the piston is at the bottom we have
$$P1V1 = 15 \text{ lb/in}^2 * 102 \text{in}^3 = 1530 \text{ lb-in}.$$
That value must be the same at the top. Therefore, $12*x = 1530$. Solving for x gives us
$$1530/12 = 127.5.$$
The new pressure at the top is 127.5 PSI. If the volume of V2 were only one cubic inch, the pressure becomes 1530 PSI. If we keep making the denominator (V2) smaller, the pressure becomes explosive. Say V2 is only 0.001 in^3.
$$\text{Then } 1530/0.001 = 1530000.$$
That's when the pressure jumps up to 1,530,000 pounds per square inch and should wear a sign saying, STAND BACK. While not practical, this is a good example of what happens

to sound when it gets squeezed to a point such that compression can no longer take place. It's almost like dividing by a number very close to zero. There is no more volume; therefore, the pressure is humongous.

The subject of a shockwave and its properties seem to defy all logic. Some shockwaves generated by explosions can travel more than twenty times the speed of sound depending on the product of detonation. A sonic boom is the remains of an airplane's shockwave as it moves from subsonic to transonic to supersonic speed. The wave front leaves the source moving much faster than 342 mps and quickly degrades as it spreads out, and then it slows to mach 1.0 by the time it reaches a person's ears.

A sequence of an F-18 Hornet going from engine idle to supersonic should help in understanding this phenomenon.

The Hornet below is at idle sending out audio waves at various frequencies. Our interest is the one with a frequency of 100Hz. It turns out that five cycles of a 100Hz wave is the length of an F/A-18/C Hornet, roughly 17.1 meters.

We watch as this band gets squashed from 17.1 meters to zero.

The airplane begins to taxi with only a small change in wave length spreading in all directions, but we are only interested in those at its nose. The length of the five cycles mimicking the piston stroke is pretty well understood; however, we must stretch our imagination to mimic the circular area to justify the volume. Imagine a cylinder with diameter of 4.66 meters. That's the height of our Hornet, so the area of the cylinder is $\pi*(4.66/2)^2$ m with a length of 17.1 m: total volume is 291.65 m^3. The idea is to imagine what happens when all the energy in a container of this size is compressed to a volume near zero. While some air is compressed in this exercise, we are only interested in sound, acoustic energy. What happens to sound when it's compressed beyond a certain limit? You can bet there will be an explosion.

In reality, there are many more cycles than five, many more Hertz than 100, and many more diameters than 4.66 meters; consequently, the concussion is huge. The length and height is in meters, so the results will be in kilo Pascals, and one atmosphere at sea level equals 101.325 kPa. Let's get underway.

The fighter begins its journey with the five cycles equal to the plane's length. Virtual volume, or V1 of our sonic cylinder is $291.65 m^3$: pressure, or P1 is 101.325kPa giving
$$P1V1 = 101.325*291.65 = 29551.436.$$

As the volume changes in the future, this number must remain the same. That means as the volume shrinks the pressure must increase.

During climbout at mach 0.5, the series is compressed from 17.10m to 8.55m; new P2=P1V1/V2, new virtual volume is 145.83 m^3. P2 has doubled at
$$29551.436 / 145.83 = 202.65 kPa.$$

Below, as the jet gains altitude its speed reaches mach 0.75, and the five cycles are now 4.275m long. Without showing the steps, pressure has doubled again at 405.29kPa.

At Mach 0.875, the five cycles have shrunk to 2.137 m, pressure at 810.58 kPa.

Notice the powers of two at work for the inverse once more. When volume is cut in half, pressure doubles.

This thing is just getting wound up at mach 0.937 with length of the five cycles now at only 1.0687m and pressure at 1621.16kPa

Then:

Chapter 22

Refraction—more factors of two

At first thought, associating flux vectors with how light reacts with various mediums may seem like a problem when refraction, reflection, and diffraction are considered. Any optic science student knows that the bending of light depends on the medium's refraction index. Several people spent many years of their lives learning how this works. They would make various types of glass and then test each type until they found their ideal refraction index.

Light speed through any solid or liquid depends on the medium's makeup, especially its boundary, how close together the molecules are, and how many electrons, protons, and neutrons make up each atom. Knowing all that, the index can be calculated. When talking about the curvature of a lens or a telescope, the result must be right on; otherwise, it would be a costly mistake. Of course, there is always trial and error.

Fermat's principle for refraction says that light follows the route of shortest time. How light determines the proper route that results in the shortest time, has always bewildered many people, especially this author. The problem includes time in the following tracks: the distance in air light must travel, and then it must bend just the right amount at the interface to come up with the correct remaining distance that results in the beam arriving at the exit interface in the shortest time. We are not including the time from the exit to the target.

Our demonstration will show only two paths; from source to interface through air, and then from interface through glass to target at exit.

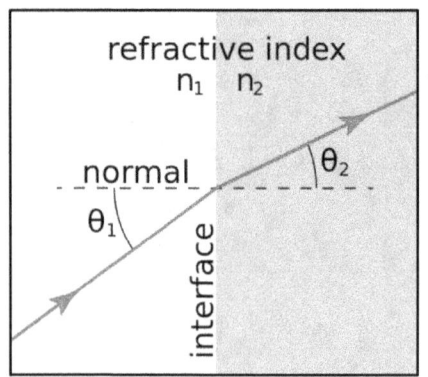

A graphical demonstration may help in understanding the relation. The drawing on the left was copied from en.wikipedia.org, a favorite reference of this author where n_1 and n_2 represent the medium's refraction index, and θ_1 and θ_2 represent angle of incidence and angle of refraction relative to the normal in that order.

Usually n_1 is air and n_2 glass or the slower medium. For the following demonstrations, the index of air is rounded down to 1.0000.

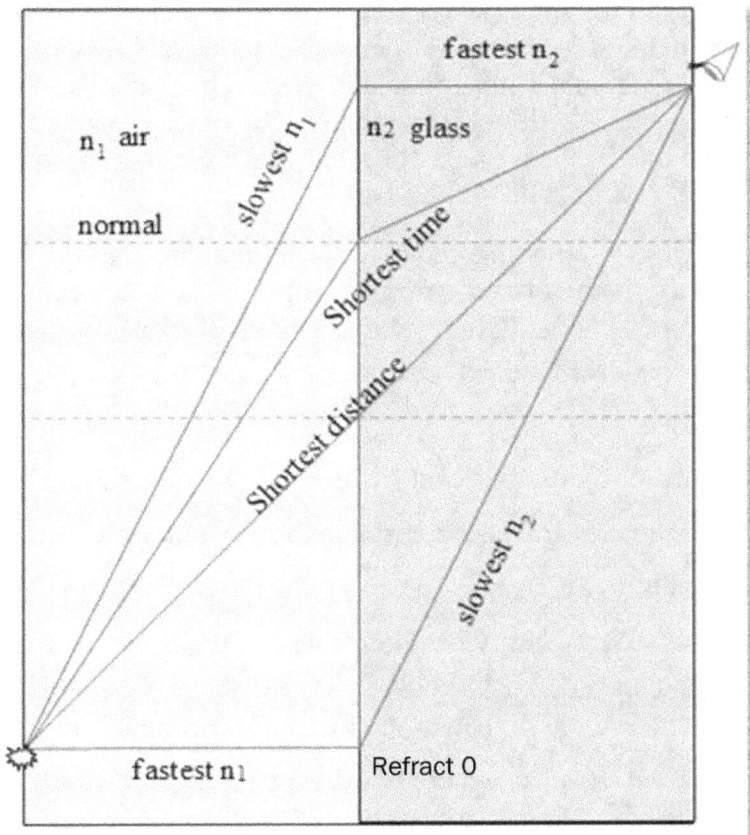

The image on the left describes four paths for a light beam to travel from its source to the detector. There are four what-if paths used to calculate the time en route.

If it goes by way of the fastest time through n_1, air, it travels perpendicular to n_2, glass, and then it bends to take the slowest time to the target. Total time in route is $fastest_1 + slowest_2 = $ longest time.

If it goes by way of the shortest distance between points without bending, it still remains in the glass much longer than in air although the distances are the same. The time en route is quicker than the previous but still too slow.

If it goes by way of the slowest time in air, then it spends too much time in air even though it takes the fastest path in glass. Then there is the compromise, the fastest total route, the ideal time in air followed by the ideal time in glass. $Ideal_1 + ideal_2 = $ shortest time.

By the way, this chapter assumes the reader is familiar with the Pythagorean Theorem. If none of this makes sense or becomes too frustrating, skip to the summary. You have done well so far, and this is the last chapter. There is no sense going nuts over these calculations. However, if you can make it through and gain the understanding, it will be worth every moment. Besides, it's possible to grasp the general idea without referring to the calculations at all.

We are dealing with four triangles in diagram Refract 1 on the following page. Triangles are in pairs with sides identified by xyz and abc. Each triangle is further identified by subscripts $_1$ and $_2$; $_1$ for air, $_2$ for glass. To make things easier, the following and its subscripts are all equal: $a = x = 1$.

Universe's Mathematics

Before we begin, instead of playing, 'Find the secret word,' we need to play, 'Find the two pair of triangles.' There are two such pairs of xyz triangles denoted by the aforementioned subscripts $_1$ and $_2$. Sub$_1$ is for triangle in air, sub$_2$ is for triangle in glass. There are also two pair of abc triangles denoted the same way. It will not be repeated. The bases are a and x, heights are b and y; hypotenuses are c and z. Please look for them in Refract 1 to become accustom to the layouts.

Beginning with the shortest path triangles xyz, the time required for light to travel along the z path will be calculated. The idea is to get the distance from the light burst to the eye by adding the hypotenuse of both triangles. Using the Pythagorean Theorem, the total distance from source to detector for triangle xyz is found by setting $z_1 = \sqrt{x_1^2 + y_1^2}$ and $z_2 = \sqrt{x_2^2 + y_2^2}$, and then total distance = $z_1 + z_2$.

Knowing the distance through each medium, time is found by dividing the distance by the speed of light. That is, $t = \dfrac{z}{c}$ where t = time to traverse z, z = length of hypotenuse in meters, and c = speed of light through medium.

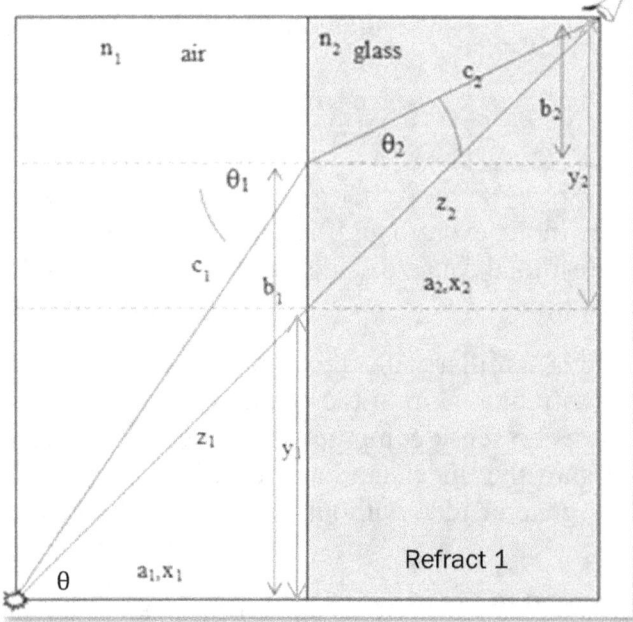

Refract 1

We're interested in the speed of light through both media, so the updated equation for time becomes $t = \dfrac{z}{c_n}$ where n is the index of the medium.

First, we calculate the time for the shortest distance, and then we calculate a longer distance with a shorter time.

To make things less complex, the speed of light through this glass is ½ that of air.

Both the air and the glass have the same dimensions of 1X2 meters. Instead of dividing distance by 300000000m to get the time with each calculation, we're going to find out how long it takes light to traverse one meter in air and glass. We do that by dividing one meter by 300000000m/s. That is, 1m/300000000m/s = 0.0000000033333s. So it only takes 3.3333 nanoseconds for light to travel one meter in air. For glass, its speed is 1/2 that of air giving 150,000,000 meters per second. 1m/150000000m/s = 0.0000000066666s;

therefore, it only takes 6.6666 nanoseconds for light to move from one end of a glass meter stick to the other.

Engineers in optics go about the task of finding refraction angles and such much differently than how this experiment will be accomplished, but our objective is to demonstrate *why* light acts as it does during refraction rather than *what* it does.

The first case is to show how much time light takes while going from source to destination by way of the shortest distance. By inspection, one can see the shortest distance is a straight line from source point to detect point. This experiment will also show how light determines the quickest time.

Remembering that a = x = base = 1, we eliminate redefining those where page size is an issue. Be aware that angles θ_1 and θ_2 are used for identification of the same angle in both xyz and abc to keep in context with Snell's Law. Further, while the computations are for θ on the air side, we use congruence of triangles to show that $\theta = \theta_1$

Time for shortest distance (air): triangle x_1 y_1 z_1	Time for shortest distance (glass): triangle x_2 y_2 z_2	Total time for shortest distance
hypotenuse is z_1: $z_1 = \sqrt{1^2 + 1^2}$ $z_1 = 1.4142$m	hypotenuse is z_2: $z_2 = \sqrt{1^2 + 1^2}$ $z_2 = 1.4142$m	
time in air: 1.4142 * 3.3333 = 4.7139ns	time in glass: 1.4142 * 6.6666 = 9.4279ns	4.7139 + 9.4279 = 14.1418ns
arc tan 1.0000 (y_1) = 45.0000° (θ_1) sin θ_1 = .7071	arc tan 1.0000 (y_2) = 45.0000° (θ_2) sin θ_2 = .7071	Check using Snell's Law $\frac{1.0000}{2.0000} = 0.5000$ $\frac{.7071}{.7071} = 1.0000$
0.5000 << 1.0000 ∴ shortest distance fails Snell's Law by far		

Table 1.1

Table 1.1 shows the calculations for the shortest distance from source (little explosion) to detector (eye) for the two triangles noted as xyz.

Snell's Law : $\frac{n_1}{n_2} = \frac{sin\theta_2}{sin\theta_1}$

Before we can use Snell's Law (above) to determine if either of these is the proper route that light takes, we must determine angles of θ_1 and θ_2.

Since we only have tangents of the angles where y_1 = 1.0000m and y_2 = 1.0000m, we use arc tan to derive both angles as in row 4 of table 1.1. We can see in row 5 that the shortest distance fails the test big time.

Table 1.2 shows the calculations for the other route using the two triangles notated as abc. That route also fails the test.

Time for next route (air): triangle $a_1 b_1 c_1$ where $b_1 = 1.5$m	Time for next route (glass): triangle $a_2 b_2 c_2$ where $b_2 = 0.5$m	Total time for next route
hypotenuse is c_1: $c_1 = \sqrt{1.0^2 + 1.5^2}$ $c_1 = 1.8028$m	hypotenuse is c_2: $c_2 = \sqrt{1^2 + .5^2}$ $c_2 = 1.1180$m	
time in air: $1.8028 * 3.3333 =$ 6.0093ns	time in glass: $1.1180 * 6.6666 =$ 7.4533ns	$6.0093 + 7.4533 =$ 13.4626ns
		Check using Snell's Law
arc tan 1.5 (b1) = $56.3099°$ (θ_1) $\sin \theta_1 = .8320$	arc tan .5 (b2) = $26.5650°$ (θ_2) $\sin \theta_2 = .4472$	$\dfrac{1.0000}{2.0000} = 0.5000$ $\dfrac{0.4472}{0.8320} = 0.5375$
$0.5000 \ll 0.5375 \therefore$ fails Snell's Law test.		

Table 1.2

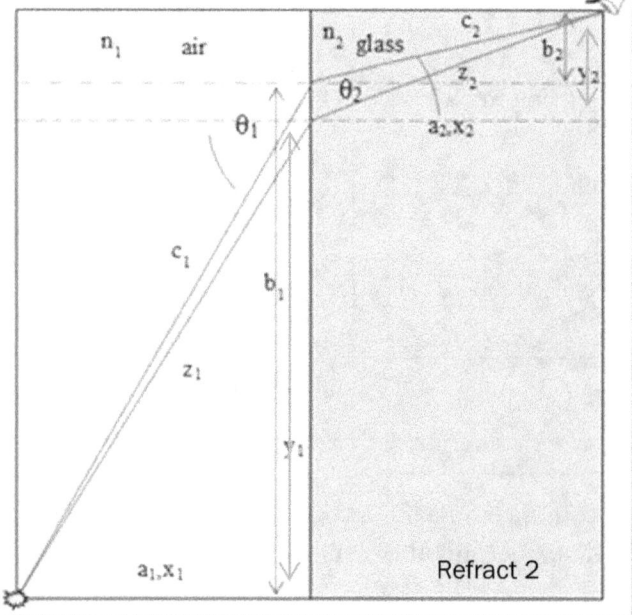

Refract 2

The next attempt is taken for the triangles of Refract 2. We will dispense with tables for the next calculations and show inline results. We begin at b_1 with a height of 1.8750m. Using the same base as before, the hypotenuse is c_1 with a length of 2.1250m. Light travel is 7.0832ns. Length of c_2 on the glass side is 1.0078m with a time of 6.7186ns. Total time is 13.8018ns.

So far, there are three different routes for light to take to its destination. The three times are: 14.1419ns, 13.4627ns, and 13.8018ns, the second being the fastest. Somehow light went from slower to faster then back to slower. Somewhere between is the ideal time.

Since abc is still too slow, backup a little and try xyz. Height of y_1 is 1.8125m resulting in $z_1 = 2.0701$m giving a time of 6.9003ns. Glass side for z_2 is 1.0174m giving a time of 6.7826ns. Total time is 6.7826ns + 6.9003ns = 13.6829ns. It's faster than the last attempt, so we're going in the right direction. But it is still longer than the minimum of 13.4627ns given earlier.

Another try on layout Refract 3: $y_1 = 1.7500$ meters resulting in $z_1 = 2.0156$m. Time to traverse z_1 is 6.7186ns. The length of z_2 on the glass side is 1.0131m. Time to traverse is 6.7539ns. Add 6.7186ns + 6.7539ns = 13.4725ns. We're getting closer to the minimum result of 13.4627ns earlier, but not quite.

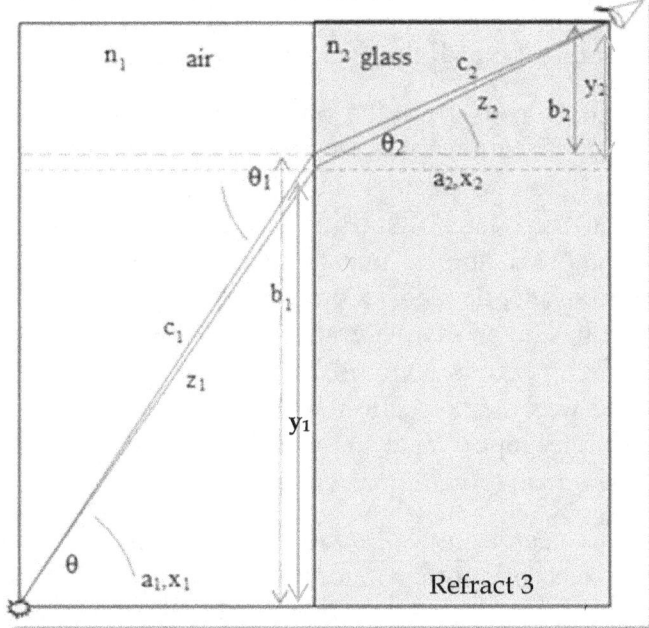

Refract 3

One more attempt before screaming like a person gone mad. The height of b_1 is 1.5384m yielding c_1 as 1.8348m. Time is 6.1161ns. Height of b_2 is 0.4616m resulting in c_2 being 1.1014. Time is 7.3426. Total time is 7.3426 + 6.1161 = 13.4587ns, the fastest yet.

Finally, we can use Snell's Law again to validate the correct angles of incidence and refraction.

As before, we are looking for angles whose tangent is 1.5384 (θ_1) and 0.4616 (θ_2). Arc tan θ_1 = 56.975°; arc tan θ_2 = 24.777°. All that is left to verify is the arithmetic.

$\dfrac{\sin 56.975}{\sin 24.777} = \dfrac{0.4191}{0.8384} = 0.4999$. And since $\dfrac{n_1}{n_2} = \dfrac{1}{2} = 0.5000$, we will call the experiment done. It's close enough for *Jazz* as some folks may say.

However, one more image will enlighten the subject further.

The image on Refract 4, next page, was generated using JavaScript to compute and draw 200 possible paths of the previous metrics. The height of sides y_1 and y_2 varies from 1 through 2 meters in .01m increments. Since the base remains at 1m, the hypotenuse of the angles in air varies from 1 thru 2.2361m while those in glass work backward from 2.2361 thru 1m.

Beginning on the left side of the image, the tracks requiring the most time are spent in glass. At attempt 154, the time reaches a minimum then begins to increase. The saw tooth means some time amplitude has been removed.

We have come to the why light takes the quickest time. The why of course is to get to its goal in the shortest time. The first to arrive at the destination is the sharpest focus; the clearest perception. All other arrivals add fussiness or color distortion.

Light emulates a shotgun blast. One buckshot will beat all others, tie with a few, and all other arrivals will enter the same hole or tear new ones.

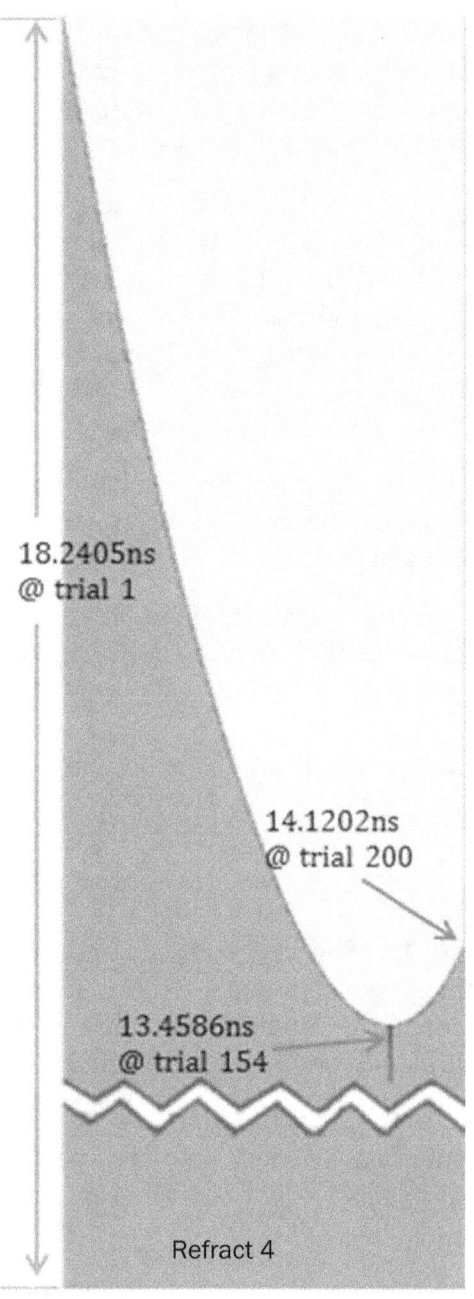

18.2405ns @ trial 1

14.1202ns @ trial 200

13.4586ns @ trial 154

Refract 4

For example, take all the trials in our experiments.

Light bursts away from the tiny explosion in the images in all directions; however, we only followed six. The last one calculated reached the target first, all others arrived later distorting things, some minor, some major. Examples may be chromatic or spherical aberrations, depending on the situation. Over the years, it has taken many iterations to correct for these late arriving beams. One of the greatest examples in our time was the correction astronauts placed on the Hubble telescope. What an amazing feat performed by everyone involved.

When the vector on which a light ray travels contacts the interface of another medium, the light ray jumps from its approach vector to another for transport through the glass. What is not shown is the exit process from the glass back to air. If the exit is at a different angle than entry, colors remain detached and go separate ways. Prisms, for example, are very interesting for anyone to explore, but we're done. Hopefully, you have made it through *Our Universe: An Alternate View* and have come out with a different perspective of your Universe than when you read the first page.

Summary

When a storm cloud expands too much to retain its configuration, its vapor changes form, cell by cell, to a more stable state—a liquid. All the energy converts to matter, cell by cell, at the same time. That is, all vapor within the cloud transforms into millions of raindrops in the same instant. This process emulates the procedure that brings forth space and matter: Water vapor simulates composite energy where each rain drop mimics each piece of matter. While rain turns into puddles, lakes, rivers and oceans, matter turns into horseshoes, horses, people and the universe.

In the beginning, there is only a web of composite energy or super energy. That composition holds all of our Universe's four super forces and all of its energy in one huge spherical globe. It is billions of light years across before a small portion of all that energy precipitates to become solid energy—matter. The matter is in the form of a filament, and its creation delivers the first cell of space. However, one of the super forces remains attached to the composite field.

And like a storm cloud that has expanded so much that its vapor changes, the whole cloud of super energy begins to condense. All of the composite energy transforms into cells of space and matter—all cells, all at the same time with the aforementioned exception.

The super force that did not transform into matter is a flux, and it remains attached to the same web from which the other three forces detached. It applies a force to the new matter and keeps super energy from crashing in on the new filament.

Nature stores all the energy within the filum, and in a short time, the cells of space begin to merge, conjoin, and work with its neighbors. With each new cell creation, radiation given off by the work done during the change of state spreads in every direction. In several billion years, it will be known as CRB.

Filaments go on to build elementary particles of matter: quarks, gluons, leptons, and other basic building blocks of yet to come about, atoms. Leptons settle down to become electrons while quarks go on to build protons, and that's when magic appears—hydrogen. Each cell of space and matter continue to merge with its neighbors to become mini-universes, and with each merge, filaments find friends they can work with. That's when things begin to go right for our young Universe.

Filums that do not form into something capable of working with others can do nothing but lie around. They are abortive sub-elementary particles or just unsuccessful subparticles depending on the stage of failure. Yet, they are still reservoirs of energy lying around awaiting an assignment. That task will be to create black holes, quasars, and control soon-to-be colossal galaxies. In 13.9 million years, these failed sub-elementary particles will be known as dark matter.

In a short time, hydrogen grows into super globes millions of earth miles across. The only remaining superforce acts on these atoms by forcing them together. Most become pairs. They work well together, and they are stable as the new universe decrees them to be. But the superforce continues to apply pressure, and as electrons are forced inwardly, they must give off energy in the form of radiation. The heat has nowhere to go except to its neighbor. Shortly this humongous sphere of hydrogen explodes generating a few lighter elements. And in the same volume previously occupied by this colossal sphere of hydrogen, several children of this cloud form stars.

Worthless globs of matter far outnumber performing matter: matter that has reached a stage of being able to perform fabulous tasks. The creation of this useless matter has also presented much more space than would have been available, and this extra size is used to separate giant configurations of hydrogen. It, along with particles and antiparticles that disappear in explosions leaving their space intact, gives our Universe 5 ¼ times more room to operate. There is more volume to arrange for greater distance between hydrogen and newly formed atoms to guarantee a successful, collective universe.

Newly formed matter at the universe's edge bring information about the boundary. That border moves away from the universe's center at the speed of light, it recedes revealing new matter daily. Most of that new matter will not make anything of value, but some of it will become bundles of stars. Those bundles of stars will form up around giant clumps of dark matter to become quasars, and then they will evolve into galaxies as time marches on. Every million or so years, new humans will find those formations very interesting, and the edge will continue to recede and create new stars and such.

Since our Universe is pressurized by the superforce, that force has a tendency to gather all matter within and place it at the center of the universe. And the center is everyplace. All objects are transparent to this flux to a degree. Flux vectors that make their way through an object leave a shadow in the background compared to oncoming flux. That shadow causes a differential force and objects move toward the less pressure. A transparency index can range from 0.000000 for a filament through 0.999999 for a gas cloud. Only a filament has no transparency. It is completely opaque; a black hole is very close to being completely opaque.

When an object is very dark, it blocks flux vectors from exiting the object and causes other vectors to bend. If no flux gets through to counter the incoming vectors, the side vectors have more influence and create a spiral effect that amplifies flux. All flux that would normally bypass the object spirals onto its surface with more crushing power.

When a runaway condition occurs, the event horizon continues to grow in an exponential manner intensifying the force on the surface in that same exponential fashion—a black hole is born.

The speed of light does not depend on electromagnetic energy itself. The waves travel on the same vectors that pressurize our Universe. The speed of superflux is just under 300,000,000 km/s. Flux vectors have an affinity for electromagnetic energy and conducts them as a copper wire conducts electricity, except the transmission is in the form of radiation.

Most people believe that mass keeps an object from traveling at c; however, that's not true. Flux vectors max out at 299,792,458 km/s, and nothing can outrun the speed of its motivating force. Actually, it has been shown that a significant drop-off in an object's trailing force begins somewhere near 10% of its maximum speed. At 0.60 c, trailing force decreases to 80% of its normal pressure. At 0.86602 c, the trailing force drops off to 50%. That's why more energy is required to advance an object in speed as it approaches that of light—it must overcome that drop-off of the aiding force.

During conduction of these virtual tests, an important concept came to fruition—space-time is a constant. That value is always one (1). At any speed, the percent of length change multiplied by the percent of time period change is always one. When traveling through space at 86.6025% c, length decreases by 0.5 and time slows by 2.0. Just for kicks, multiplying it through we get

$$0.5 * 2.0 = 1.0$$

Whereas our Universe is expanding, it is doubtful that it is growing near the speed of light. Space and matter were born through expansion, and that gave impetuous to what continues today. However, the fact that something twice the distance away is traveling at twice the speed of nearby objects can be misleading. Distance and speed are taken in series and must be summed up as such. Molecules at the end of a red-hot poker travel faster than those in the middle because they are at the end of a string of molecules in series. The rate of expansion of each particle is the same for an identical temperature, and when they are summed together, the end particles must move at the total rate of all those between. That information is usually in a machinist's or engineering handbook. But there is no handbook for the universe.

What is the universe expanding into? Whatever it is, the universe must be replacing it with itself. It cannot be space because that is part of the universe. It can't be nothing because nothing is only a word that has no equivalency to anything at all. Yet, it is expanding. Let's think about it for a moment.

What marks the boundary of our Universe? It is the CRB, and on the other side of that border lies composite energy: energy from which everything came. It applies a restriction to expansion of the young universe, and our solar system is now in the same place as that composite energy was billions of years ago—right smack dab in the center! Furthermore, each part of our solar system is at the center of its own universe.

So, as composite energy expands, three of the four super forces can no longer sustain the same form. The remaining superforce stands fast creating space and time while the other three create matter by instilling all energies and forces into this new state, and it continues today just as it has been doing so for 13.9 billion years.

So, when was the beginning of time? Time begin at the same instance space was created. Space has two properties: length and time. We could just as well refer to space by itself with the understanding that its length-time is always present. When one increases, the other decreases proportionally and vice versa.

· · · · ·

A side note: Father Georges Lemaitre could still look back to a time when the universe was much smaller and much younger the same way he did in the early 20th century because this view of the beginning doesn't change a thing. It just changes the origin of our Universe from an explosion of nothing to a transformation of a Grand Energy to space and matter.

By-the-way, there are a couple more methods that could produce differential forces on matter, for now it is transparency index or shadows.

Appendix A

Vectors

This is a vector. Name it **a**. It has both direction and magnitude.

The arrow or head points in its direction; its length is its magnitude.

This is another vector. ⟶ Call it **B**. It is different than **a** in both magnitude and direction.

Usually a vector name is in bold face type.

Another way to identify a vector is tail to head where   A is tail and B is head.

But those characters only identify the beginning and end. To identify the vector itself the head and tail must be barred. That is, vector $\overline{AB}$.

We can refer to **a** either as   or as

We can also add vector **a** to **B**.   Or **B** to **a**,   where

Resultant is equal to **a** + **B** or **B** + **a**.

We can also subtract them.   **a** - **B** or **B** - **a**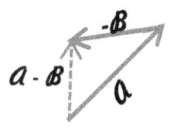

163

Now come the calculations.

Each vector has two Cartesian coordinates, x and y.

Notice each vector represents part of a triangle, the hypotenuse, and it has both x and y components.

For example, take *a* from above and enlarge it for better visuals, we get

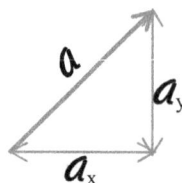

If we know its x, y values we can find the magnitude of *a* by calling on Pythagoras again. That is, $x^2 + y^2 = a^2$. Or $a = \sqrt{x^2 + y^2}$. Say x is 19 and y is 19 mm. That results in $\sqrt{361 + 361} = \sqrt{722}$. Giving *a* as 26.87 mm long.

Okay. That's fine for a single vector, but we need to add two or more as above. What now?

That brings on another way of describing *a*. The new way is to introduce its coordinates,

a(x, y). From above we have *a*(19, 19), and that gets us one vector. The next thing is to grab the (x, y) values of the guy to be operated on. Say it's *B* also from above. Since its y value is small compared to its x, it also requires enlargement.

Now that the values are known, it is simply addition from here on.

a(19, 19), *B*(23, 3.43). Adding the x components give 19 + 23 = 42 mm. Adding the y components give 19 + 3.43 = 22.43 mm.

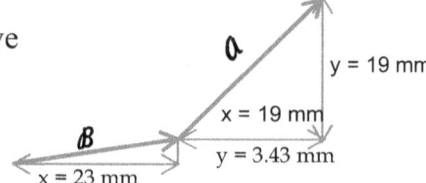

Incorporating the new values results in a new triangle, *C*, with the base equal to the sum of the x components and a height equal to the sum of the y components.

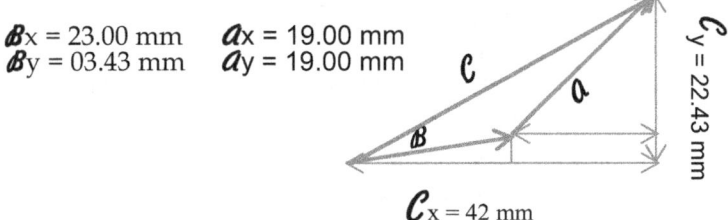

Still, there is something missing. Visibly, we recognize *C* as a vector because its direction is implied by its arrow tip pointing in the northeasterly direction. The only value recorded is its magnitude. To prove it is a vector without visual aid, the direction must also have a value.

This is where trigonometry helps, but superfluous information must be cleared away first. Since $y/x = \tan \theta$, we can just ask whose tangent of θ is $22.43/42.00 = 0.5340$. Otherwise known as $\arctan(0.5340) = 28.10°$. So *C* has a magnitude of $\sqrt{42^2 + 22.43^2}$ which comes down to 47.61 mm in the direction of 28.10°.

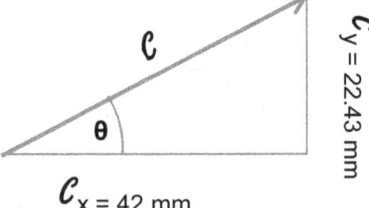

Superflux vectors are added in the same way.

Here *a* and *B* are added to determine how much force is applied to an object and in what direction.

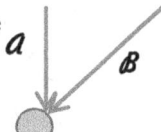

However, magnitude must include the vectors as a flux field instead of vector length.

Vectors can be added or subtracted or multiplied, and they can be multi-dimensional. For more on the subject visit a fine web site: https://www.mathsisfun.com/algebra/vectors.html

Glossary of Terms

∴ :
Symbol meaning therefore. Used in logical and mathematical results

Accelerometer:
A devise that measures the acceleration of gravity in g-units. Scale begins at 1 g

Aether:
From en.wikipedia.org
According to ancient and medieval science, aether, also spelled æther or ether and also called quintessence, is the material that fills the region of the universe above the terrestrial sphere.[2] The concept of aether was used in several theories to explain several natural phenomena, such as the traveling of light and gravity. In the late 19th century, physicists postulated that aether permeated all throughout space, providing a medium through which light could travel in a vacuum, but evidence for the presence of such a medium was not found in the Michelson–Morley experiment, and this result has been interpreted as meaning that no such luminiferous aether exists.

Affinity:
The degree to which one thing attracts or assists another

AKA:
Also Known As

Amplitude:
The maximum extent of a vibration or oscillation, measured from the position of equilibrium

Antimatter:
Any matter that has an opposite charge of the universe's accepted version of that matter

Apsis:
An extreme point of an object's orbit. Closest is perihelion. Farthest is aphelion

Apsidal precession:
Rotation of an apsis around its host. Orbital movement of an object's aphelion and perihelion around another body

Asymptotically:
Adverb for asymptote which is a line that can approach some value but never get there.

Attenuation:
The reduction of the force, effect, or value of something. The reduction of the amplitude of a signal, electric current, or other oscillation

Barycenter:
Current model: Astronomical center of gravity of a cosmic system
Alternate View model:
    Astronomical center of superflux differential forces of a cosmic system

Big Bang:
Current model: birth of universe
Alternate View Model:
    Composite energy gives birth to universe

Black hole:
Current model: Highly massive accumulation of matter where nothing can escape its gravity
Alternate View model:
    An accumulation of matter so dense no flux can make its way through leaving a solid black shadow. Flux within the event horizon spirals down to the surface amplifying the pressure many times

BYO:
Billion Years Old

Chinese puzzle:
A pair of nails in a loop such that they can be coupled and decoupled in the exact reverse manner

Composite Energy:
A pre-universe state where all matter, space, and energy are combined into a single form

Conservation of energy:
Energy can change states but can never be destroyed. Each state change is always equal to the original quantity but may be in different forms

Conjoined:
Past of conjoin. To combine as one with common use of parts

Cosmic Radiation Background:
The left over energy from the birth of the universe

Cosmoses:
Plural of cosmos which means the universe as a whole

CRB:
See Cosmic Radiation Background

Dark energy:
Current model: Mysterious energy forcing the universe to expand faster than it would without it.

Alternate View model:
A glob of composite energy that has not completed the process of changing states to space and matter

Deoxyribonucleic Acid:
Built in information giving direction to organic cellular construction.

DNA:
See Deoxyribonucleic Acid

Deuterium:
A hydrogen atom whose nucleus contains one neutron

Doppler Effect:
A change of an object's space-time according to its velocity. Denoted by a measured change of frequency emitted by transmission and reception

EWAG:
Educated Wild Ass Guess

Electromagnetic energy:
A combination of electronic and magnetic energy. Light and radio waves reveal the energy as it propagates along its way

Electromagnetic force:
One of the four fundamental forces of the universe. It exerts a force to all atomic elements but effects iron and its derivatives the most

Electron:
A negatively charged atomic particle. Spends most of its time orbiting a nucleus of some sort

Electron volt or electronvolt or eV:
A metric used for various outcomes, sometimes energy, sometimes mass, and many others.
For energy:
$1 \text{ eV} = 1.602 \times 10^{-19}$ Joules
For mass:
$eV/c^2 = 1.782662 \times 10^{-36}$ kg

Element:
From Oxford Dictionaries: Each of more than one hundred substances that cannot be chemically interconverted or broken down into simpler substances and are primary constituents of matter. Each element is distinguished by its atomic number, i.e. the number of protons in the nuclei of its atoms.

Elementary particle:
Current model: The smallest unit of matter that can no longer be broken into something smaller.
Alternate View model:
An elementary particle is built of filaments

EME:
Acronym for Electromagnetic energy.
See Electromagnetic energy

Ex-energy:
All matter used to be part of composite energy; therefore, all matter is ex-energy

Failed subparticles:
Subparticles that failed in becoming something capable of working with others. They are hiding in our Universe as dark matter

Fermions:
From en.wikipedia.org: In particle physics, a fermion (a name coined by Paul Dirac from the surname of Enrico Fermi) is any particle characterized by Fermi–Dirac statistics. These particles obey the Pauli exclusion principle. Fermions include all quarks and leptons, as well as any composite particle made of an odd number of these, such as all baryons and many atoms and nuclei.

Filum:
A solid form of composite energy. The smallest form of matter to exist. Latin term for filament.

Flux:
Flux may be singular as a unit vector or multiple as a flux field. For superflux it is both. In space it is a field, but when it contacts an object the field may breakup into yet smaller fields, even into a single vector. The default is movement until something stops it. Its speed of travel is 299,792,458 m/s

Flux goggles:
An imagined set of glasses with a special filter for observing oncoming superflux.

Flux shadow:
See transparency shadow

Fulcrum:
The pivot point of a leaver or balance beam

Gamma:
Gamma (γ) relating to the Lorentz factor defined as $\gamma = \dfrac{1}{\sqrt{1-\frac{v^2}{c^2}}}$

Googolplex:
One google is $1*10^{100}$.
One googolplex is $1*10^{google}$ or $1*10^{10^{100}}$

Inflation:
Just after the Big Bang there were no laws governing speeds or anything else. According to Professor Alan Guth (MIT) energy of the bang spread at a rapid rate across the new universe.

Isotope:
An atomic element with any number of neutrons in its nucleus. The behavior remains the same, but the weight increases dramatically.

Kuiper Belt:
From solarsystem.nasa.gov:
Both 2014 MU69 and Pluto are in the Kuiper Belt. It is a disc-shaped region beyond Neptune that extends from about 30 to 55 astronomical units (compared to Earth which is one astronomical unit, or AU, from the Sun). Comets from the Kuiper Belt, known as short period comets, take less than 200 years to orbit the Sun and travel approximately in the plane in which most of the planets orbit the Sun. There may be hundreds of thousands of icy bodies and a trillion or more comets in this distant region of our solar system.

Lepton:
An elementary particle of two classes—charged and non-charged. The charged class consists of the electron

Magnetic field:
A vectored field that describes the magnetic influence of electrical currents and magnetized materials. Gauss is its unit of measure

Magnetic flux:
See Magnetic field

Matter-Antimatter collision:
Two separately charged subparticles come into contact and all their combined matter becomes energy

Megasecond:
One million seconds or about 11.57 days

Milankovitch cycle:
From Wikipedia:
Milankovitch cycles describe the collective effects of changes in the Earth's movements on its

climate over thousands of years. The term is named for Serbian geophysicist and astronomer Milutin Milanković.

Mini universe or mini-universe:
One of trillions of small universes at birth just after creation of space and matter

Mole: abbr mol
From Wikipedia:
It is defined as the amount of a chemical substance that contains as many elementary entities, e.g., atoms, molecules, ions, electrons, or photons, as there are atoms in 12 grams of carbon-12 (^{12}C).
This number is expressed by the Avogadro constant, which has a value of $6.022140857 \times 10^{23}$

Momentum: id is ρ (Greek letter rho):
A built in property of an object, a state of motion of an object at rest or moving:
ρ = mv (mass * velocity); it keeps on doing what it's doing until something interferes.

Nanoseconds:
$1/1,000,000,000^{th}$ of a second or one billionth of a second

Neutrino:
Little dude sent through the universe when a neutron spits out its electron

Neutron Star:
Ultra-heavy star consisting mostly of neutrons; billions of neutrons can fill the place of one hydrogen atom

Node:
A connecting point or intersection of space cell networks

Nodule:
A group of nodes

Normal:
A line that is perpendicular to another line or reference. Used to make a right angle

Nucleus:
The central core of an atom containing most of its material of protons and neutrons

Opaque:
Degree of which light can pass through an object. 100% blocks all light, 0% blocks none

Orbital:
Description or identifier of an electron's orbit about its nucleus. Usually n?

Periapsis:
The point of an orbit where the planet is nearest the host

Precession:
Movement of a rotating body about an axis when an external torque is applied.

Proton:
A subatomic particle with a positive charge that is 1836 times heavier than an electron

Quark:
An elementary particle that makes up complex subatomic particles like protons and neutrons

Quasar:
QSO (quasi-stellar object) or star-like object with a black hole emitting a tremendous amount of energy. Now considered to be at the center of most galaxies

Radiation, Electromagnetic:
A kind of radiation including visible light, radio waves, gamma rays, and X-rays, in which electric and magnetic fields vary simultaneously.

Receding edge:
The universe is growing at light speed. The edge is always at the end of the radius, so as the edge gets farther away, it is known as receding. It is revealing new information daily.

Redshift:
A change in frequency of light toward the red end of the spectrum. Initiated by the emitting object's movement away from a detector.

Resonance:
A suitable frequency of vibration that is ideal for a given length of material that produces the greatest amplitude with less dampening.

Space:
An entity that contains the only remaining force that did not separate from composite energy. That remaining force keeps the universe from collapsing by applying pressure throughout its space

Space-Time Constant:
All objects in motion reduce distance in the direction of travel and stretch distance in trail. A change in distance brings on a change in time. The product of distance and its time period must always equal one (1).

Space-Time:
Two factors of the universe related to orientation. Space without time does not exist. Time without space does not exist. They affect each other. They are something like electromagnetic energy which contains both electric and magnetic energy.

Speed of Light:
Light travels on superflux at the constant rate of 299,792,458 meters per second usually rounded up to 300 million m/s. The space-time constant guarantees this rate.

String Theory:
In physics, string theory is a theoretical framework in which point-like particles of particle physics are replaced by one-dimensional objects called strings. It describes how these strings propagate through space and interact with each other. From *en.wikipedia.org*

Strong force:
See strong interaction force

Strong Interaction Force:
It is the strongest of the four fundamental nuclear forces, 1) electromagnetic force, 2) gravitational force, 3) weak force, and 4) the strong force. However, Alternate View believes gravity did not break away.

Subatomic particle:
Accepted definition: A subatomic particle is made of elementary particles

Alternate View: Filaments make up currently known elementary particles

Sub particle or subparticle:
See subatomic particle

Super Energy:
The parent of the currently known four fundamental forces. Part of composite energy

Superforce:
What is currently known as gravitational force did not separate. It remained attached to the fabric of space-time to keep composite energy from crashing in on matter

SWAG:
Simple Wild Ass Guess

Sway:
Encouragement to believe some idea. Persuade.

TBD:
To be determined

Tidal Forces:
A differential pressure applied to various parts of an object attempting to separate them from the whole

To be determined:
Not yet fully rationalized or designed

Transparency:
A degree of opaqueness in a negated sense

Transparency index:
A scale one or less quantitizing the degree of transparency

Transparency shadow:
A shade produced by the background of superflux

Transparency voids:
A term used to describe a transparency shadow. When the shadow becomes too dark other flux vectors tend to change directions to fill the missing vectors

Tritium:
: A hydrogen atom whose nucleus contains two neutrons

Vector:
: A single line of flux that has direction

Versin:
: An older trigonometrical function used in celestial navigation. Versin $\theta = 1 - \cos \theta$

WAG:
: Wild Ass Guess

Weak Force:
: See weak interaction force

Weak interaction force:
: The weakest of the four fundamental nuclear forces

Index

acceleration, 70, 71, 72, 73, 74, 75, 76, 77, 78, 80, 81, 82, 88, 94, 97
accelerometer, 71
affinity, 161
AKA, 24, 31
Alan Guth, 2
antimatter, 6
atom, 1, 7, 9, 14, 16, 32, 54, 70, 104, 152
barycenter, 78, 79, 81, 82, 142, 143, 144
Big Bang, 1
black hole, 13, 17, 31, 35, 52, 55
bosons, 7
bulkhead, 96, 97
BYO, 28
Chinese puzzle, 5
Christian Doppler, 64
composite energy, 1
conservation of energy, 5
Cosmic Radiation Background, 6
CRB, 18
dark energy, 10
Deoxyribonucleic Acid, 7
deuterium, 9
DNA, 7
Doppler Effect, 64
electromagnetic energy, 7
electromagnetic force, 1
electron, 6, 86
electron volts, 8
element, 9
elementary particles, 7
EME, 83, 167
ex-energy, 7
fermions, 7
filaments, 2
filums, 2, 9, 13, 39
Flux, 3
flux goggles, 47
flux shadow, 37
flux shadows, 42, 74
flux transparency shadow, 35
fulcrum, 78
gamma, 94
gravitational force, 1
helium, 16
Hydrogen, 9
Inflation, 2
isotope, 9
Kuiper belt, 54
lepton, 6
leptons, 6, 7, 8, 26
magnetic field, 7
magnetic flux, 84
matter-antimatter collisions, 13
microsecond, 96
Milankovitch cycle, 140
mini universe, 20
moles, 114
mols, 114
momentum, 5, 81
Momentum, 5, 77
nanoseconds, 75
neutrino, 9
neutron, 9
neutron star, 58
newtons, 88
nodes, 7
nodule, 7
nuclear fusion, 16
nucleus, 9
opaque, 31
orbital, 16
precession, 55
proton, 8
quarks, 7
Quasar, 55
radiation, 6, 21
Receding Edge, 26
repulsion, 6
Resonance, 5
semi-void, 49
Space, 1
Space-Ti
Space-Ti
speed of
String The
strong for
strong inte
sub atomic
sub particle
subparticles
super energ
superforce,
TBD, 8
Tidal forces,
to be determi
TOC, iii
transparency,
transparency s
transparency v
tritium, 9
vector, 35
versin, 78
void, 51, 52
weak force, 7
weak interaction
μsec, 97

Tritium:
A hydrogen atom whose nucleus contains two neutrons

Vector:
A single line of flux that has direction

Versin:
An older trigonometrical function used in celestial navigation. Versin $\theta = 1 - \cos \theta$

WAG:
Wild Ass Guess

Weak Force:
See weak interaction force

Weak interaction force:
The weakest of the four fundamental nuclear forces

Index

acceleration, 70, 71, 72, 73, 74, 75, 76, 77, 78, 80, 81, 82, 88, 94, 97

accelerometer, 71

affinity, 161

AKA, 24, 31

Alan Guth, 2

antimatter, 6

atom, 1, 7, 9, 14, 16, 32, 54, 70, 104, 152

barycenter, 78, 79, 81, 82, 142, 143, 144

Big Bang, 1

black hole, 13, 17, 31, 35, 52, 55

bosons, 7

bulkhead, 96, 97

BYO, 28

Chinese puzzle, 5

Christian Doppler, 64

composite energy, 1

conservation of energy, 5

Cosmic Radiation Background, 6

CRB, 18

dark energy, 10

Deoxyribonucleic Acid, 7

deuterium, 9

DNA, 7

Doppler Effect, 64

electromagnetic energy, 7

electromagnetic force, 1

electron, 6, 86

electron volts, 8

element, 9

elementary particles, 7

EME, 83, 167

ex-energy, 7

fermions, 7

filaments, 2

filums, 2, 9, 13, 39

Flux, 3

flux goggles, 47

flux shadow, 37

flux shadows, 42, 74

flux transparency shadow, 35

fulcrum, 78

gamma, 94

gravitational force, 1

helium, 16

Hydrogen, 9

Inflation, 2

isotope, 9

Kuiper belt, 54

lepton, 6

leptons, 6, 7, 8, 26

magnetic field, 7

magnetic flux, 84

matter-antimatter collisions, 13

microsecond, 96
Milankovitch cycle, 140
mini universe, 20
moles, 114
mols, 114
momentum, 5, 81
Momentum, 5, 77
nanoseconds, 75
neutrino, 9
neutron, 9
neutron star, 58
newtons, 88
nodes, 7
nodule, 7
nuclear fusion, 16
nucleus, 9
opaque, 31
orbital, 16
precession, 55
proton, 8
quarks, 7
Quasar, 55
radiation, 6, 21
Receding Edge, 26
repulsion, 6
Resonance, 5
semi-void, 49

Space, 1
Space-Time, 8
Space-Time Constant, 88
speed of light, 12
String Theory, 2
strong force, 7
strong interaction force, 1
sub atomic particle, 9
sub particle, 9
subparticles, 14
super energy, 1
superforce, 31
TBD, 8
Tidal forces, 44
to be determined, 8
TOC, iii
transparency, 31
transparency shadow, 44
transparency voids, 49
tritium, 9
vector, 35
versin, 78
void, 51, 52
weak force, 7
weak interaction force, 1
μsec, 97

www.ingramcontent.com/pod-product-compliance
Lightning Source LLC
Chambersburg PA
CBHW062215220526
45471CB00009B/3208